Process Improvement with CMMI® v1.2 and ISO Standards

Process Improvement with CMMI® v1.2 and ISO Standards

Boris Mutafelija
Harvey Stromberg

CRC Press is an imprint of the
Taylor & Francis Group, an **informa** business
AN AUERBACH BOOK

Auerbach Publications
Taylor & Francis Group
6000 Broken Sound Parkway NW, Suite 300
Boca Raton, FL 33487-2742

Auerbach is an imprint of Taylor & Francis Group, an Informa business

Printed in the United States of America on acid-free paper
10 9 8 7 6 5 4 3 2 1

International Standard Book Number-13: 978-1-4200-5283-1 (Hardcover)

Library of Congress Cataloging-in-Publication Data

Mutafelija, Boris.
Process improvement with CMMI v1.2 and ISO standards / Boris Mutafelija, Harvey Stromberg.
p. cm.
Includes bibliographical references and index.
ISBN 978-1-4200-5283-1 (hardcover : alk. paper)
1. Quality control--Standards. 2. ISO 9001 Standard. 3. Capability maturity model (Computer software) I. Stromberg, Harvey. II. Title.

TS156.M8746 2008
658.5'62--dc22 2008040988

Visit the Taylor & Francis Web site at
http://www.taylorandfrancis.com

and the Auerbach Web site at
http://www.auerbach-publications.com

Contents

Dedication

To our wives Mirta and Susan and our ever-growing families:

our children, their spouses and significant others.

Thanks for your understanding and support—

Without your love and patience this book would not have been possible.

Authors

Boris Mutafelija is Principal Member of Technical Staff at Systems and Software Consortium, Inc. He has over 35 years of information technology experience as an engineer, software professional, and manager. Mutafelija led several organizations in reaching higher process maturity levels (as defined by the SEI). He developed process architectures, worked on establishing process frameworks for efficient process improvement, and taught, tutored, and consulted many teams in process improvement.

Mutafelija's process improvement interests include process frameworks, enterprise aspects of process engineering and improvement, measurements, statistical process control, and, of course, using multiple standards in developing effective and efficient process improvement approaches. With Harvey Stromberg, he coauthored the book, *Systematic Process Improvement Using ISO 9001:2000 and CMMI*, coauthored over 30 papers, and is the coinventor of three U.S. patents. Mutafelija is authorized by the SEI as a lead appraiser for performing SCAMPI appraisals and as an instructor for delivering Introduction to CMMI courses.

Harvey Stromberg is with BAE Systems. He has over 35 years experience in systems and software engineering, quality assurance, and process improvement in diverse industries. He has managed development projects, quality assurance and configuration management departments, and engineering process groups. In those positions and as a consultant, he has helped bring several organizations to higher CMM/CMMI maturity levels and transition from ISO 9001:1994 to ISO 9001:2000.

Stromberg's process improvement interests include the use of standards when developing effective and efficient process improvement approaches and the application of measures for process and project management. He is the coauthor of *Systematic Process Improvement Using ISO 9001:2000 and CMMI* and is an authorized instructor for the SEI Introduction to CMMI course. He is also a senior member of the American Society for Quality (ASQ), a Certified Quality Manager (CQM), and a Certified Software Quality Engineer (CSQE).

Acknowledgments

After we wrote our first book, many standards—including CMMI—changed. Readers would stop us at conferences to ask if we were going to write a revised edition to accommodate those revisions. In the meantime, we worked with several organizations that used more than one framework and saw the need to say something more about that. We decided that in addition to ISO 9001 and CMMI, we needed to consider several of the most "popular" ISO standards: ISO 20000 as well as the two lifecycle standards, ISO 12207 and ISO 15288, which have recently been revised and harmonized.

There are too many people to mention whom, in many ways, helped us with this book—knowingly and unknowingly. But first and foremost, there is one person who "infected" us with the writing bug. He had faith in us when we decided to write our first book and continued with his encouragement when we decided to write this one, even when we had to postpone the delivery because the most current revisions of the standards were not yet available. This person is John Wyzalek, senior acquisition editor, Auerbach Publications, Taylor & Francis Group. Thanks, John!

Over the years we had the opportunity to interact with the management and staff of the Software Engineering Institute—they were always ready with advice and encouragement. Many thanks!

Also, we wish to express our gratitude to our colleagues at the Systems and Software Consortium, Inc., who enabled us to experiment with multiple frameworks and share our approaches with member companies.

And a special note: while writing this book, Boris became a grandpa to twin girls, Grace and Julia, who, with their tiny presence, provided such great love and satisfaction that the effort for finishing this book was made much less difficult.

We are aware that our maps are subjective and imperfect, and that some of our colleagues will have differences with our approach. We encourage all to provide feedback and, to those who do so, we thank you in advance.

Boris Mutafelija
Harvey Stromberg

Foreword

Boris and Harvey have done it again. The CMMI Product Team—including experts from industry, government, and the Software Engineering Institute (SEI)—has continued to evolve the CMMI Product Suite to make it relevant to a broader audience, now reaching version 1.2. These two authors have similarly evolved their fine first book to address more of the relevant additional models that organizations need to investigate to cover their commitment to quality and process improvement—all in their most current versions.

With the Product Team, we first added clearer hardware coverage in the development "baseline" (CMMI-DEV). Then, given the architectural adjustments made as we began the version 1.2 redesign, we began the construction of two variants to enter the closely associated areas of acquisition of products and services from suppliers (CMMI-ACQ) and of service delivery (CMMI-SVC). (For those of you who may not have fully transitioned from v1.1 to v1.2 of CMMI, *Process Improvement with CMMI® v1.2 and ISO Standards* contains an appendix to assist your understanding of those changes as well as the main tasks of relating CMMI to the ISO world.) Boris and Harvey have matched our efforts by adding new standards to their collection. In each case, they map these to CMMI. We, at the SEI, appreciate this, as we are often asked to relate CMMI to one of the other key standards that have been effectively created by the various ISO committees and writing teams. In addition to the critically important development-related standards for software engineering (ISO 12207) and systems engineering (ISO 15288), they have added the services standard (ISO 20000) as well as ISO 90003. In addition to supporting the process improvement journey, they have sought to clarify the many areas where a single piece of objective evidence can indicate the performance of the practices captured in multiple models and standards. This synergy reduces the total cost of appraising progress against each of the standards and models.

In my foreword to their 2003 book, I expressed my appreciation in that Boris and Harvey maintained the emphasis on process improvement rather than favoring achieving maturity levels or ISO certification. The current environment encourages me to reemphasize this priority. We have been disappointed to find that some organizations, after devoting sufficient attention to the best development practices contained in CMMI, have failed to deliver the performance on some defense programs that would reasonably be expected to be achievable. Although in many cases the complexity of the development and less than "mature" acquisition guidance may have significantly contributed to the shortfalls, there has too often been tendencies to "check the box." This book continues the intended use of the best practices of the various standards and models—to assist the creation of high quality products in the most timely, economic fashion.

Within the last year, we have conducted workshops to elicit ideas about future evolution of the CMMI Product Suite. One of the strong interests expressed was to enhance the ways that CMMI could be used with other improvement approaches and the relevant international standards and various bodies of knowledge. We have committed a portion of our 2009 budget to explore options and are encouraging sponsorship from industry for ways to leverage the ideas and improvements. Boris and Harvey are providing invaluable leadership with this book to fuel the move forward. Your use of the tools and techniques captured within this work will enable you to join us in the effort to mature these ideas—and improve your individual approaches as we move forward together. The benefits continue for all of us.

Mike Phillips
CMMI Program Manager

Chapter 1

Introduction

Companies that survive, thrive, and grow are constantly changing and improving. Change is often easiest to implement when improvements are guided by existing roadmaps or standards. This sounds straightforward in principle, but there are so many standards out there! Selecting the right standard can be baffling and the number of choices leads to many questions, among them:

- Which standard should be selected?
- What if more than one standard seems to fit the company's needs?
- What if no single standard addresses all needs?
- What should be done when some standards address our business needs but others are required by customers?
- If multiple standards are selected, should any one of them be chosen as the primary driver for process definition and improvement?

In many businesses, an organization's compliance to the requirements of a given standard will be formally evaluated and measured. Achievement of a specified level of compliance is often required by customers and may be an important driver for process improvement. Two prominent frameworks frequently used for compliance assessments are ISO 9001:2000, for quality management, and CMMI v1.2, for systems and software engineering process improvement. The ISO 20000:2005 standard, with its strong relationship to the Information Technology Infrastructure Library (ITIL), is also emerging as a frequently used compliance framework. Another category of standards is composed of structured collections of best practices. Among the standards in that category are:

- ISO 15288, for defining systems engineering life cycle processes
- ISO 12207, for defining software life cycle processes
- ISO 90003, for applying ISO 9001 to software

In fact, grouping standards this way (compliance frameworks vs. best practices collections) is not strictly accurate. Although ISO 9001 and CMMI are frequently selected for formal assessments, both are valuable as sources of best practices. Similarly, the standards listed as collections of best practices can serve as the reference models in assessments. Overlaps among standards are inevitable because they were developed independently, under different schedules, and had different sponsoring groups. Even though those overlaps may pose problems during implementation, they exhibit synergy. It is this synergy that we explore in this book. This book will show how to implement, manage, and sustain process improvement, guided by an integrated application of the appropriate standards.

We will present some of the widely used fundamental concepts of process improvement including discussion of enablers and barriers to process improvement, highlights of popular process improvement approaches, and topics to consider when using multiple frameworks. The main features and purpose of each standard discussed in this book are summarized. We also discuss the changes from CMMI v1.1 to CMMI v1.2 and efforts to align and harmonize multiple standards.

We then bring the multiple elements together, showing how to exploit the similarities and differences among the standards and capitalize on their synergy. We also discuss how the objective evidence needed to measure conformance to multiple standards may be efficiently created, organized, and presented.

Finally, we show the multiple relationships among the standards at a detailed requirement level. In general, these relationships, or maps, are "many-to-many," meaning that a requirement in one standard may correspond to more than one requirement in another standard, and vice versa. Also, the detailed relationships vary in their strength or degree of correspondence.

These relationships become important both when developing and implementing a process improvement strategy and during appraisals. Because the real needs of the business must be the drivers for improvement, the domains of each framework, their areas of overlap, and their differing approaches must be understood so the frameworks can be effectively and efficiently integrated. An organization's quality management system or set of standard processes must be designed to serve the needs of the organization and not merely to address the requirements of one or more standards.

Audits and appraisals are expensive and time consuming. An organization seeking certification or appraisal against more than one framework can realize significant returns by minimizing the effort needed to collect the objective evidence used for each evaluation event.

Although previous process improvement experience is helpful, it isn't necessary to benefit from reading this book. Specifically, this book will be beneficial to the following groups:

- *Process improvement practitioners* who must develop strategies for process improvement implementation and the transition to new or revised standards—Process improvement practitioners develop and support the processes that will be implemented and institutionalized. They need to have full knowledge of the requirements of multiple standards so they can identify the processes that should be selected for improvement.
- *Evaluators* making compliance decisions and recommendations—Evaluators compare the actual implementation of processes and practices to standards and judge the degree of compliance. They need to understand the interactions among standards when developing findings and making recommendations.
- *Senior managers* making decisions on standards selection and implementation—Senior management provides leadership, resources, and funding for process improvement and standards implementation. An understanding of the principles on which each standard is founded and the synergy among standards is valuable for their decision making.
- *Process implementers* applying the processes that have been developed to support the selected standards—Understanding the features of each standard provides the knowledge that will help them understand why and how the organization's processes are defined and provide guidance for collecting objective evidence needed for appraisals.

The frameworks addressed in this book represent multiple viewpoints and are based on many years of experience in different domains. They overlap in some places and address unique niches in others. Taken together they can provide a valuable foundation for real process and quality improvement.

The ISO 9001 standard defines the requirements for quality management systems and is broadly applicable to many domains. The other frameworks we discuss, however, are aimed at specific problem areas, namely, systems engineering, software engineering, and IT service management. Their domains and a qualitative view of their areas of overlap are shown in Figure 1.1. Specific details illustrating their relationships are discussed later in this book, notably in chapters 5 and 6.

These frameworks adopt a variety of approaches to presentation and differ in the level of detail provided. In some cases, notably in CMMI, substantial amounts of informative material and many examples are provided. For other frameworks, such as ISO 9001, the presentation is rather skeletal and interpretation is largely left to the implementing organization (and auditors). The range of detail presented in each framework may be inferred by examining the number of pages of text provided in each framework, as shown in Figure 1.2.

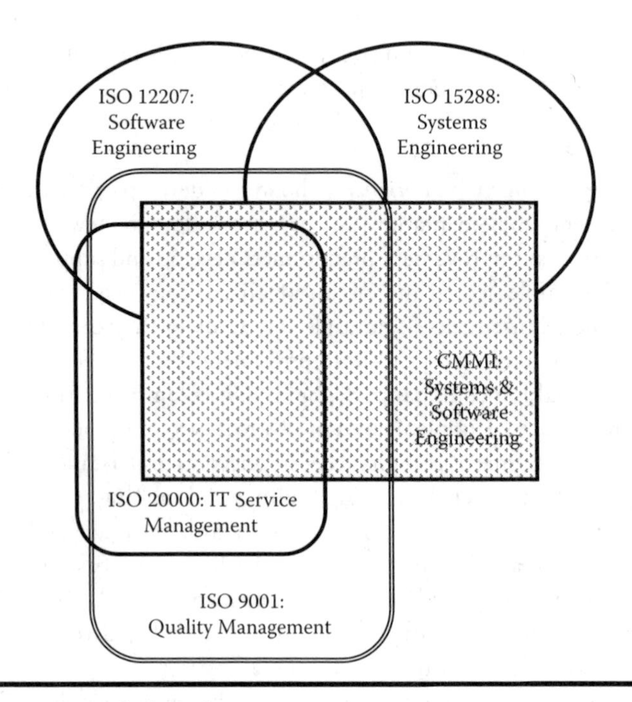

Figure 1.1 Framework domains.

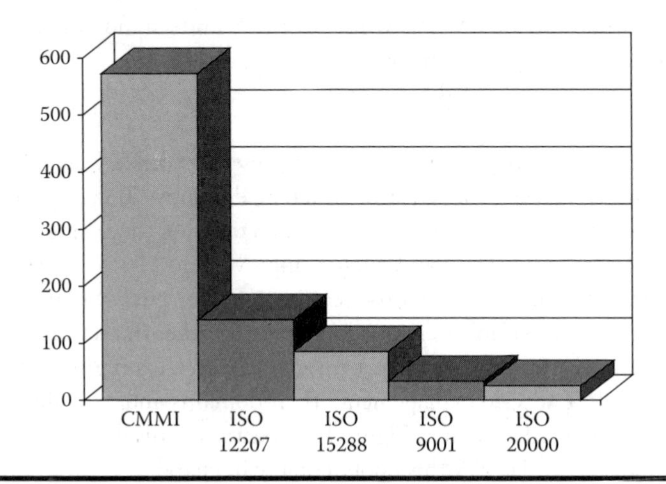

Figure 1.2 Framework size.

Structural differences also distinguish the frameworks. The process areas and practices in CMMI include the concepts of increasing capability or maturity. These concepts may be applied to individual process areas or to predefined groups of process areas. Within limits, organizations may choose the sequence in which they wish to address the process areas and the targeted capability level. In contrast, the selected ISO standards define the requirements for processes, activities, and tasks but do not describe levels of achievement.

As we will see when we discuss appraisals, the time and effort needed to implement processes that are compliant with these frameworks varies considerably. This is due, in part, to the degree of emphasis placed on institutionalization and, in part, to the type and amount of objective evidence expected by each appraisal or certification method.

Clearly, an organization has many things to consider before embarking on process improvement efforts using multiple frameworks. Among the topics to consider are:

- Should we adopt any framework at all? Organizations with well-established processes and infrastructure may decide that adoption of a framework or standard may not bring meaningful business benefits.
- Is certification or appraisal against a framework a mandatory business requirement? In many industries, certification against one or more standards is a de facto entry criterion for doing business.
- Should multiple frameworks be used? Are different frameworks required for different customers? Is a single framework inadequate for addressing the needs of the organization's products and services?
- If multiple frameworks are used, should one of them be the main driver for the organization's processes? If, for example, ISO 9001 certification is needed for certain customers, perhaps the organization's processes should focus on addressing that standard before adding practices driven by a different framework.
- Should the same set of frameworks and organizational processes be used across the entire enterprise? Are the same capability or maturity levels appropriate across the enterprise?

Let us note that while the focus of this book is on exploiting the synergy of multiple frameworks for process improvement, another approach is to integrate several frameworks into a single model. One noteworthy example is CMMI itself, which is largely based on three predecessor capability maturity models. Examples of other efforts include:

- FAA-iCMM—Version 2.0 of this framework, developed by the U.S. Federal Aviation Administration, was released in 2001. The framework addresses acquisition, supply, engineering, development, operation, evolution, and support.
- Integrated System Framework (ISF) for Excellence—This framework is being developed by Integrated System Diagnostics (ISD) and incorporates several of the frameworks discussed in this book as well as other models, such as the People CMM, and improvement concepts, such as Six Sigma.

Finally, we should mention that there are other ongoing efforts to develop maps between models that may use different mapping paradigms and sometimes come to different conclusions. Such efforts include work by the Software Engineering Institute (SEI) summarized in the 2007 draft report (Kitson et al. 2007) where CMMI v1.2 is compared to ISO 9001. Although this draft report doesn't address multiple frameworks, it provides a detailed mapping between those two major frameworks.

The book is organized as follows:

- Chapter 1—This introduction.
- Chapter 2—This chapter discusses the motivations for process improvement, the factors that make improvement difficult, and the conditions and activities that help enable improvement. We address two widely used process improvement approaches, namely, Plan–Do–Check–Act and IDEAL[SM]. The advantages and issues associated with the use of multiple frameworks are introduced.
- Chapter 3—The structure and content of Capability Maturity Model Integration, version 1.2, are summarized here. We discuss the characteristics and expectations of each process area (PA). Several viewpoints are represented here as the PAs are examined by category, by lower- and higher- maturity groupings, and by staged and continuous representation.

The concept of "constellations," which is new to version 1.2, is introduced. The changes from CMMI v1.1 are also summarized. These changes provided clarifications, reduced model complexity and size, and enabled broader model coverage.

- Chapter 4—This chapter provides summaries of the ISO standards addressed in this book. Our intention here is to provide enough information to allow the reader to use the summaries as a guide to the standards and as an aid to understanding the more detailed maps between frameworks presented in subsequent chapters and appendices. It is not intended as a substitute for the details and nuances to be found in the standards themselves. The standards discussed are:
  - ISO 9001:2000, a broadly applicable quality management standard
  - ISO 90003:2004, guidance for interpreting ISO 9001 for software
  - ISO 15288:2008, defining life cycle processes for systems engineering
  - ISO 12207:2008, integrated with ISO 15288 and defining processes for software engineering
  - ISO 20000:2005, providing standards for information technology service management
- Chapter 5—Having presented summaries of the frameworks of interest, we now get down to the business of discussing the detailed relationships or "maps" between four ISO standards and CMMI. (Because ISO 90003 provides guidance and interpretation, but does not add requirements to ISO 9001, their maps to CMMI are identical.) Our approach to mapping and

some of the considerations and trade-offs involved in developing maps are presented.

- Chapter 6—We can now attempt to bring all the elements together and discuss the synergy and use of multiple frameworks. We take a CMMI-centric view and add the other frameworks. The approach, however, works just as well if an organization starts with a different framework.
- Chapter 7—Finally, we consider the need to conduct appraisals against multiple frameworks and the need to efficiently collect the objective evidence required by each standard and appraisal method. Similarities and differences among the appraisal and certification approaches are discussed.
- Appendices—The appendices provide additional detailed information as follows:
  - A—Acronyms
  - B—References
  - C—Changes from CMMI v1.1 to CMMI v1.2
  - D—ISO 9001 to CMMI map
  - E—ISO 15288 to CMMI map
  - F—ISO 12207 to CMMI map
  - G—ISO 20000 to CMMI map

For the reader's convenience, detailed maps and corresponding correlation matrices are available for downloading from the publisher's Web site, http://www.crcpress.com/e_products/downloads/download.asp?cat_no=AU5283. This will enable the users to further experiment with the framework relationships. In the spirit of true process improvement, we look forward to obtaining comments on those maps and matrices.

Chapter 2

Process Improvement Fundamentals

This chapter presents some of the fundamental concepts of process improvement including defining the motivation for change, identifying process improvement barriers and enablers, and selecting improvement approaches. An overview of the Plan–Do–Check–Act (PDCA) approach generally favored by ISO standards and the SEI's IDEAL^{SM} approach is provided. The use of the IDEAL model as a process improvement roadmap is presented and concepts associated with using multiple frameworks are introduced.

Introduction

Process improvement efforts are started for many reasons. Traditionally, the primary goal of process improvement has been to improve efficiency. Improved efficiency, or productivity, allows an organization to produce the same products with reduced effort. Frederick Taylor's (1911) work in *The Principles of Scientific Management* was focused on selecting workers for the particular physical activity that best suited them, training them in the optimal methods for performing their tasks, and ensuring that those scientifically developed methods were followed. This approach led to monotonous work, few opportunities for feedback, and little chance to change the way things were done.

Fortunately, we've made quite a bit of progress since then.

Some process improvement goals have been with us for a long time—improving cost and schedule performance and improving product quality are always important.

"Faster, better, cheaper" is an oft-heard mantra. Organizations also look for better predictability; whether performance is good or poor, organizations need to be able to rely on repeatable performance so that accurate plans and budgets may be developed and commitments may be made.

Companies often succeed because they have highly skilled people, effective tools and technology, and are in advantageous business environments. Process definition and improvement is a mechanism for keeping these components in balance. More important, when situations change, as surely they will, robust processes provide the mechanism for adaptation to those new situations (Garcia and Turner 2006).

Quality Improvement

W. Edwards Deming (1982) described the "quality chain reaction": When product and service quality improves, costs decrease because there is less rework and fewer delays. Because costs have decreased, productivity is improved. Higher productivity lets a company capture a bigger piece of the market with lower price and higher quality. Greater market share leads to more business and higher employment. Thus, a focus on quality leads to increased business success. Process definition and process improvement enables this desired improvement in product and service quality.

Having effective, efficient, and predictable processes in place does more than merely improve an organization's business performance. The ability to produce and deliver products and services at the promised cost, schedule, and quality levels also leads to satisfied customers. Furthermore, in an environment where continuous improvement is a way of life, employee satisfaction increases. Employees not only know what is expected from themselves and others, but they also have a way to provide their feedback and make improvements. They are not automatons, but are thinking and valuable contributors. This sense of self-worth helps reduce employee turnover and attracts others to the organization.

Dealing with Multiple Frameworks

Over the years, numerous quality management standards, engineering standards, and maturity models have been developed. We refer collectively to these documents, which provide the structure for capturing concepts, practices, and relationships, as frameworks. Frameworks include both models and standards. Each framework typically includes training material, interpretation guidance, and audit or appraisal approaches. The frameworks often address similar topics, but present information in different ways, addressing different disciplines, and using different language. Even where frameworks are each primarily focused on different topics, there is often some degree of overlap.

Organizations may adopt certain frameworks because they have a need to display appraisal results or a registration certificate. ISO 9001 registration or the achievement of a particular CMMI maturity level is frequently a business prerequisite. Even when these achievements are not strict requirements, an organization with those credentials will often be perceived to deliver higher quality products and thus have an advantage over other organizations.

Organizations may voluntarily choose to use more than one framework or may be driven to adopt those frameworks to meet statutory or contractual requirements. In either case, a comprehensive process definition and improvement effort is generally needed to eliminate redundancies, overlaps, and conflicts. In addition to eliminating duplication and inconsistencies among the frameworks, a process improvement program based on multiple frameworks may be used to:

- Enhance the understanding of interactions among business and technical functions
- Develop common components suitable for use in several technical disciplines
- Reduce the cost of implementing process improvements when several disciplines have to be considered

The use of multiple frameworks, however, involves a greater effort and expense than the use of a single framework. The requirements of each framework must be considered in conjunction with the other frameworks in use. The vocabularies of each framework may be different and overlapping areas and conflicts must be addressed. Organizations undergoing appraisals and certification for several different frameworks may find themselves collecting the same evidence multiple times and burdening their staff with nondevelopmental work. Furthermore, frameworks and their supporting components evolve and are revised independently, so keeping the relationships among requirements up to date is difficult, even within a related "family" of standards.

Impediments to Change

As Watts Humphrey, founder of the Software Process Program of the Software Engineering Institute (SEI) at Carnegie Mellon University, has said: "Process improvement may be simple, but it isn't easy." Most people and their organizations find it difficult to make changes in the way they operate. Even where the current processes don't work very well, they're still "our processes" and there's a level of comfort in continuing the status quo.

One of the most common pitfalls for process improvement programs is the failure to realize how much time and effort it takes to define and deploy process changes. While improvements are being made, the staff is still under day-to-day pressure to deliver products and services. Even with the guidance of a process improvement framework to help identify common topics and activities, organizations need to spend time determining:

- Business goals and objectives
- The current baseline(s)
- The improvement target(s)
- Which changes are the most important

Care must be taken to avoid addressing too many changes concurrently or in too short a period of time. Every organization has a different culture; some organizations can welcome change more easily than others, but none have unlimited capacities. Changing one set of processes will typically affect other concurrent and interacting processes, so the ripple effect must be considered.

If management sets improvement programs in motion but does not actively demonstrate sustained support or provide adequate staff or budget, the risk of failure increases. Each failed improvement initiative makes the next one more difficult to implement. Without a doubt, those members of the organization who most fear losing influence and power will be sure to publicize earlier difficulties. Thus, the history of prior improvement initiatives should be considered when improvement activities are planned.

Probably the biggest impediment to implementing changes arises when an organization focuses on achieving an appraisal score or certification instead of acting to make real improvement. The certificates can be won—usually at great expense—but there will only be the appearance of change. The organization's staff will have learned to pay lip service to the improvement activities and will have learned that real change isn't supported.

Process Improvement Enablers

Fortunately, many years of experience across many industries have helped determine the characteristics of successful process improvement efforts.

If we don't know where we are going, we'll probably never get there. Or, as Watts Humphrey put it more eloquently, "If you don't know where you are, a map won't help." An organization must understand its improvement goals, the barriers to success, and the activities needed to reach its goals. An organization with process improvement goals tied to business objectives is more likely to succeed than one with an amorphous improvement program or one tied only to a certification target.

Process improvement goals must be realistically achievable and must be founded on an accurate knowledge of the current state of process definition and maturity. Because all goals cannot be addressed at once, the number and scope of improvement goals addressed must be commensurate with the time and resources allocated.

Performance can realistically be expected to improve incrementally, but a leap from a low maturity state to world-class status is unlikely. Small improvements can accumulate and lead to long-term gains. In fact, one enabler of process improvement success is process improvement success! Although the first efforts may be difficult to define and implement, effective improvement helps build a culture that is open to change.

One consistently cited theme in successful process improvement programs is the value of support and commitment from upper and middle management. Of course, it's critical for the process improvement initiative to be adequately funded and resourced, but that isn't enough. Continuing management interest in the status and active participation in the progress of improvement efforts makes all the difference. Higher-level management can show its support and continuing interest by reinforcing the implementation of new policies and procedures and by making sure that program reviews include discussion of process improvement status and implementation of appropriate actions. Management support must continue even during periods of stress, which may cause priorities to shift.

Higher-level management must also support middle management's efforts to improve processes. In the abstract, nobody at any level is opposed to improving quality and productivity. In reality, middle managers are often evaluated on cost and schedule performance to a far greater extent than they are evaluated on quality and process improvement performance. Middle managers must be allowed to pursue improvements without being squeezed on budgetary and milestone status above all other considerations.

Process improvement efforts typically produce new processes, procedures, work instructions, templates, training, and deployment activities. These artifacts and activities must be aligned with the work and culture of the organization. The collection of processes that is appropriate for a large U.S. Department of Defense contractor will not satisfy the needs of a small commercial organization. Neither can the processes used by a high-maturity division of a company be effectively transplanted to a low-maturity division. When the breadth and depth of process definition appropriately matches the organization's needs, the probability of success is increased.

One approach to implementing appropriate and effective processes is to involve the people who will have to use the processes in developing those processes. Full-time process improvement specialists may have extensive backgrounds, a deep understanding of the relevant frameworks and standards, and, certainly, the best of intentions, but if they are not actually implementing those processes their insights will be limited. This issue becomes particularly important when the process improvement program is based on multiple frameworks and the resultant organizational processes may be deployed across diverse domains.

Many of these enabling factors are supported through the organization's process improvement infrastructure. An illustration of a typical approach is shown in Figure 2.1. Here, a Management Steering Group (MSG) provides high-level direction

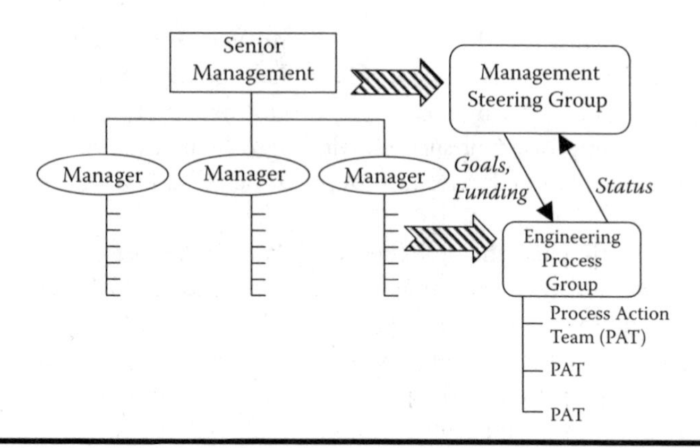

Figure 2.1 Process improvement organization structure.

and funding. The MSG membership comprises representatives of the organization's higher-level management. Such a committee has the clout to provide resources, set priorities, and remove impediments.

Day-to-day work is done by a group typically known as a Process Group (PG), Systems Engineering Process Group (SEPG), or an Engineering Process Group (EPG). The EPG is composed of both full-time process improvement professionals and practitioners who are assigned for a limited—but meaningful—period of time. Staff assigned to the EPG must make a significant time commitment. A time commitment of only a few hours per week will result in very low productivity because the EPG member will spend too much time in context switching, recalling the goals of the assigned task, and remembering the work that has already been done. An approach that is even worse than making small time commitments is assigning staff on a "time available" basis or because the staff has no other current assignment. This is sure to result in other activities taking priority over process improvement work. The EPG typically charters focused teams, variously known as Process Action Teams (PATs) or Technical Working Groups (TWGs), to address specific issues. PAT membership is generally drawn from the EPG and from other practitioners assigned for the particular task. Issues addressed by a PAT might include reporting on the current state of the practice for a selected topic, developing a new process, or revising training material. EPG and PAT members in a multiple framework environment will need to have a good understanding of the relevant frameworks and their relationships.

With a process improvement organizational infrastructure in place, process improvement activities can be run as a project. Plans can be developed, resources allocated, responsibility and authority assigned, and implementation can be monitored and managed. Thus, improvement tasks are less likely to slip between the cracks, and the risks and results of those tasks can be identified and shared.

Approaches to Implementing Change

Having identified the motivation for change, and with an awareness of process improvement pitfalls and enablers, it must still be recognized that improvement efforts will not yield an unbroken chain of successes.

Change means disrupting the way things are currently done. Disruption will initially bring confusion, uncertainty, and a loss of productivity. Until there is some degree of understanding of the changes, feedback on the results, and stabilization, the benefits of the improvements will not be realized. Worse yet, not all changes will actually be improvements. Sometimes, the ideas that seemed so wonderful on paper don't work very well. As Yogi Berra reportedly said, "In theory there is no difference between theory and practice. In practice there is."

A systematic approach to implementing change is needed so that an organization can withstand unproductive and chaotic periods and, through appropriate measures, determine how well the improvements are working. Several approaches have been developed over time. Some ISO standards such as ISO 9000 and ISO 20000, which will be discussed in chapter 4, explicitly advocate the use of the Plan–Do–Check–Act (PDCA) approach. The SEI has developed and advocates the use of the IDEAL approach. A different approach is described in ISO/IEC 15504:2004, *Information Technology – Process assessment* (ISO 2004b). Still others base improvements on variations of the spiral model (Boehm 1988).

Beyond the activities described in any of these approaches, all process improvement programs also require activities to manage the overall effort and coordinate the various tasks. Resources must be allocated and balanced against other organizational needs, activities must be prioritized, and efforts must be monitored.

We will discuss two of the best-known improvement approaches here, namely, PDCA and IDEAL.

Plan–Do–Check–Act (PDCA)

The Plan–Do–Check–Act (PDCA) cycle is also known as the Shewhart cycle or the Deming cycle. It is one of the best known approaches for implementing improvements and provides the foundation for many other methods. It is intended to be implemented repeatedly, with each cycle building on previous improvements and the knowledge that has been accumulated. In other words, it forms the basis for continuous process improvement. The PDCA cycle, illustrated in Figure 2.2, consists of the steps shown in Table 2.1.

IDEAL

IDEAL is a process improvement life cycle model developed by the SEI (McFeeley 1996). It was originally developed to help support software process improvement efforts conducted in conjunction with the CMM for Software, but has been extended

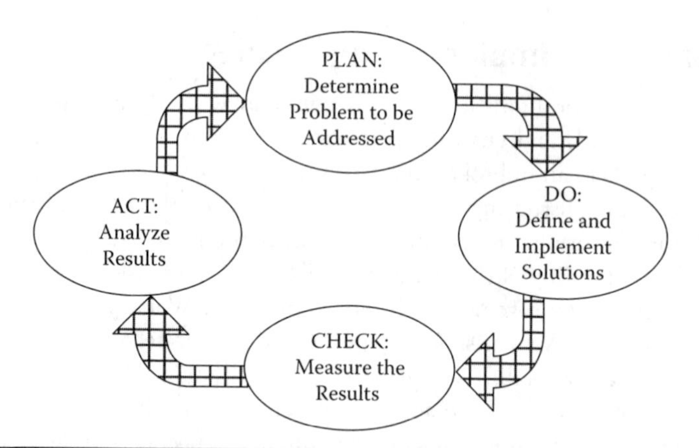

Figure 2.2 Plan–Do–Check–Act cycle.

Table 2.1 Plan–Do–Check–Act Cycle

Step	*Activities*
Plan	1. Select the problem(s) to be addressed. This may be done using a variety of tools and techniques such as flowcharting, Pareto analysis, and Ishikawa diagrams.
	2. Establish measurable improvement targets.
	3. Identify the processes affecting the problem area and analyze them to identify potential causes of problems.
	4. Collect data related to the problem and the candidate processes.
Do	1. Develop potential solutions and criteria for selecting a solution.
	2. Select a solution.
	3. Implement solution on a limited, pilot basis.
Check	Evaluate the results by collecting and analyzing measurements.
Act	1. If desired results were achieved, formalize process, develop training, plan for deployment, and monitor results. Begin the PDCA cycle again to address the next improvement.
	2. If desired results were not achieved, begin the PDCA cycle anew to identify alternative causes and solutions.

to have broader process improvement application. Each letter in the IDEAL acronym represents the name of a process improvement phase, namely:

- I—Initiating
- D—Diagnosing
- E—Establishing
- A—Acting
- L—Learning

The cyclical nature of the IDEAL model is shown in Figure 2.3. Not unlike the PDCA approach, an improvement program using the IDEAL model will make many trips around the improvement cycle. The improvement cycle is initiated in response to one or more drivers. These "stimuli for change" are generally associated with business needs, such as a need to improve quality, productivity, or predictability.

In the *initiating* phase, improvement goals, which support the overall business strategy, are selected and the expected benefits of achieving those goals are identified. In less mature organizations, benefits may be qualitative rather than quantitative. Goals and expected benefits may evolve over time, but the intent at the time of their initial selection should be made clear. The senior managers who will commit resources and provide leadership and support should be identified during this phase because their sponsorship will be needed throughout all improvement activities. The improvement infrastructure, such as one previously described comprising a steering committee, a process group, and working groups, should also be established. The organization should recognize that the composition of those infrastructure groups will most likely evolve and change as process improvements are developed and implemented.

An organization that intends to use more than one framework to guide process improvement activities must be sure to make the connection between business goals and the selected frameworks. Members of the infrastructure groups should

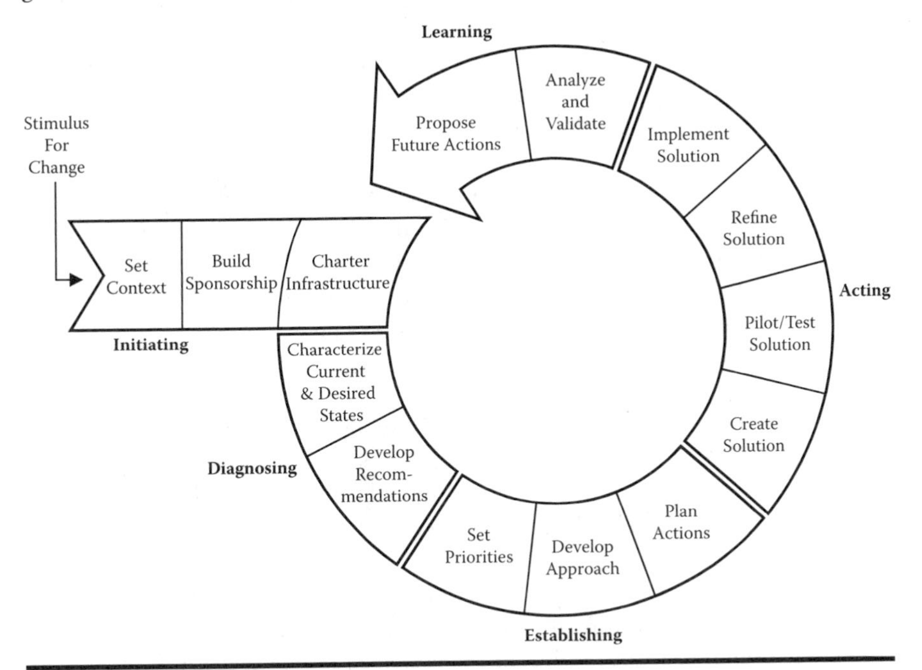

Figure 2.3 The IDEAL model.

establish an understanding of the relationships among the frameworks, the vocabularies used, the similarities, and the differences.

Because this phase involves identifying and prioritizing organizational goals and committing resources to the improvement efforts, the initiating phase activities are primarily the responsibility of higher-level management, with guidance from process improvement professionals.

The *diagnosing* phase is used to develop an understanding of the current state and define the desired state. Evaluation of the current state is generally done using a defined reference framework, such as ISO 9001 or CMMI. Definition of the desired state is driven by evaluation against one or more reference models while considering the goals initially identified as the stimuli for change. Here, it is important to understand the relationships among the selected frameworks to minimize duplication of evaluation effort and to avoid development of conflicting results.

An EPG, chartered during the initiating phase, is generally responsible for planning and executing the diagnosing phase. EPG members will need to have knowledge of the selected reference model or models, the appropriate evaluation methods, current organizational processes, and the activities actually performed by practitioners. The activities actually performed may not be addressed by current process definitions or—amazing, but true—may be in conflict with process definitions.

Work plans are identified in the *establishing* phase. Recommendations developed in the diagnosing phase are prioritized, giving consideration to dependencies, resource constraints, outside pressures, and other organizational activities. The prioritized recommendations are used to develop an overall approach, which, in turn, leads to a detailed implementation plan. The detailed plan will include responsibilities, assignments, tasks, schedules, measurements, reviews, and a risk management strategy.

Organizations using multiple frameworks must take special care to develop plans that insulate practitioners from the vagaries of each framework. Members of the process improvement infrastructure need to have a detailed understanding of the standards being used; others in the organization do not. One overall goal is to have the bulk of the staff follow the organization's processes without thinking in terms of "ISO processes" or "CMMI processes" as entities different from processes used to perform day-to-day work.

Drafting the process improvement plan is usually the responsibility of the EPG, but the MSG and middle management should be closely involved. Implementation of the plans developed in this phase will make demands on the organization's resources and may affect ongoing work, so those stakeholders need to be able to make informed commitments.

In the *acting* phase, the good intentions of the preceding phases are turned into actions. This is the most difficult and time-consuming phase of the IDEAL cycle. The implementation plans created in the establishing phase lead to the development of solutions, which are typically tested on a limited number of pilot projects.

After being tested, the solutions are adjusted and deployed across the organization. Deployment may be conducted in a variety of ways, such as:

- Implement on all projects on a just-in-time basis
- Implement incrementally on defined sets of projects
- Implement on one business unit at a time
- Implement on projects larger (or smaller) than a defined threshold value
- Combination of these above approaches

The acting phase has many “moving parts” and parallel activities. Responsibility for various tasks may be allocated as follows:

- EPG—Day-to-day management and coordination of process improvement activities, including tracking budget, schedule, and technical progress against plans, is handled by the EPG. The EPG manages deployment and measures the effect of improvements.
- PAT—The PATs develop solutions to their assigned problems, including process definitions, training, and supporting assets. Under management of the EPG, new processes may be piloted and may lead the PAT to revise or tune the solution.
- MSG—The MSG sponsors provide ongoing support and involvement in improvement efforts. The MSG monitors progress and readjusts priorities and resources as needed.
- Practitioners—Practitioners, the target audience for improved processes, receive training on new or revised processes. They implement processes and provide feedback to PATs and EPG.

The *learning* phase is used to determine and evaluate the effects of the changes that have been implemented. Here, the organization can determine if the goals that motivated the changes have been successfully addressed, analyze the lessons learned, and develop proposals for future improvements. Although this phase is depicted in Figure 2.3 as following the acting phase, it will usually begin after some improvements have been deployed and other solutions are being developed.

Generally, the EPG or PATs are responsible for reviewing the lessons learned and measuring progress toward the identified goals. Their improvement suggestions become inputs to the next cycle of process improvement.

Frameworks

Nothing in the preceding discussion of PDCA and IDEAL has focused on the use of any specific standard or framework. The PDCA and IDEAL approaches and the frameworks and standards discussed in this book are intentionally written as general guidance to be applicable in many environments.

Why, then, should an organization adopt one or more general standards to guide improvement efforts instead of developing its own organization-specific guidance? As previously noted, the use of a specified framework is often a precondition to conducting business. However, many organizations adopt these frameworks even in the absence of contractual requirements because they represent structured distillations of best practices. International standards, and de facto standards such as CMMI, are developed over many years using thousands of hours of expert analysis. The resources used to examine actual results, build models, define interfaces, and develop examples are almost always far greater than any single organization can bring to bear.

Because these standards are written to be broadly applicable, any organization can—and indeed, must—develop implementation guidance that matches the needs, priorities, and constraints of its environment.

If the use of one quality or improvement standard is good, then surely the use of several must be even better! Well . . . perhaps.

As a matter of fact, there are advantages to using more than one framework in the same improvement effort. An informal audience survey at the SEPG 2007 meeting (Siviy et al. 2007) identified several of the most important benefits arising from the synergy of using multiple models. These benefits include:

- Complementary strengths of frameworks can be exploited
- Coverage of each area is more complete
- A holistic view is provided
- A single point of view is avoided and more creative thinking is supported
- Religious wars over favorite standards are avoided
- A larger selection of tools and techniques is enabled by different perspectives
- Time and resources are more efficiently used

The use of multiple frameworks is not, however, without significant challenges. Each framework has its own sponsoring group or organization, uses its own vocabulary, is focused on satisfying a specific set of needs, and may have its own appraisal method. Harmonization of multiple standards is not often a high priority for the sponsoring groups. Another survey at SEPG 2007 highlighted the following obstacles to using multiple frameworks:

- Groups using multiple models must understand the goals, benefits, and technical details of each model
- The connections, similarities, and differences among models must be understood
- Schedules for framework changes are uncoordinated and asynchronous
- Organizational processes must change in response to framework changes

- Practitioners may resist adopting multiple models
- Senior management may see models as competing or may be interested in limiting support to a small number of models

Thus, we want to determine how these models and standards can be brought together so that the benefits of using multiple frameworks can be maximized and the challenges can be minimized. In the following chapters, we will discuss the important characteristics of CMMI and five ISO standards. We will then be ready to address approaches for tying the frameworks together and taking advantage of their synergy.

Here's an example of how the IDEAL model can be used with the material presented in this book to create a process architecture using CMMI and four ISO standards.

Initiating Phase

Based on market needs and contractual requirements, organizational management decided that it is necessary to consider CMMI and four ISO standards for process improvement. They have set a target to achieve CMMI ML 3 and, based on market intelligence, foresee that ISO 9001 and ISO 20000 compliance may be required. At the same time they learned that there may be some contractual requirements for software to follow, but not necessarily conform to, ISO 15288 and ISO 12207. Therefore, those two standards should be used to provide additional process details that may satisfy some life cycle model requirements. The Engineering Process Group (EPG) was tasked to develop a process improvement plan that would reflect those business objectives. For this purpose they considered many CMMI practices and the mappings found in chapters 5 and 6.

Diagnosing Phase

To determine the current state of standardization and process implementation, the EPG performed a series of gap analyses to benchmark existing organizational and project processes against each of the selected frameworks. Based on the material in chapter 7, they selected SCAMPI C as a benchmarking method. They decided to use the CMMI-centric approach described in that chapter and used the ISO-to-CMMI mapping when collecting benchmarking evidence. By reading the framework descriptions in chapters 3 and 4, they realized that they also needed more knowledge about each of the frameworks so they could understand the mappings and the evidence to be collected so they took several courses covering those frameworks.

Establishing Phase

In this phase, the EPG developed an improvement approach and planned their process development, implementation, and deployment activities. They were guided by CMMI practices for developing and implementing improvement plans and the framework mappings described in chapters 5 and 6 and in the appendices. Most of the EPG process improvement work was performed in this phase. A process architecture was developed and process elements to be developed were identified. They also identified those processes that did not do well during benchmarking and developed plans for improvement and for achieving compliance with one or more frameworks. They established guidelines for implementation of multiple frameworks as part of the design of process elements and the development of a process improvement strategy.

Acting Phase

In this phase the EPG executed the plans, deployed processes on the projects, and started process improvement. Guidance for this phase was provided by CMMI practices for process deployment and implementation and was supported by the mappings described in chapters 5 and 6.

Learning Phase

In the learning phase, projects and functional groups provided feedback to EPG indicating successes and barriers to process deployment and improvement. Most of this work was supported by CMMI practices addressing the incorporation of lessons learned into organizational standards and other assets, and by the mappings described in chapters 5 and 6.

Summary

The motivation for improvement has been around for a long time. However, differing business needs have led to the development of many standards and frameworks, some meant for general use and some focusing on particular domains. It is often beneficial to use multiple frameworks to take advantage of the best practices and lessons learned that they have captured.

In this chapter, we have summarized two of the best known process improvement approaches, namely, PDCA and IDEAL. Over time, common process improvement enablers as well as the typical impediments to change have been identified. We can use these lessons learned when applying systematic approaches to process improvement to take advantage of the synergy of multiple frameworks.

In chapters 3 and 4, we will provide overviews of CMMI and the selected ISO standards.

Chapter 3

Capability Maturity Model Integration (CMMI)

This chapter provides an overview of Capability Maturity Model Integration, version 1.2 (CMMI v1.2). We discuss the architecture of the model, model components, model representations, and the characteristics of the model's process areas (PAs). Changes from CMMI v1.1 and the effect of those changes on process improvement activities are summarized.

Introduction to CMMI v1.2

Background

CMMI v1.1 was published by the Software Engineering Institute (SEI) in December 2001 as several 700-page reports and later, in 2003, as a single book. The CMMI framework contains a product suite, which is based on industry best practices and incorporates lessons learned from predecessor models, primarily:

- Capability Maturity Model (CMM) for Software, v2.0, draft C (and, unofficially, CMM v1.1)
- EIA/IS 731, System Engineering Capability Model (SECM)
- Integrated Product Development Model, draft v0.98 (IPD-CMM)

Those best practices are embodied in four bodies of knowledge, or disciplines: systems engineering, software engineering, integrated product and process development, and supplier sourcing. In August 2006, CMMI v1.2 was released, incorporating lessons learned from the use of CMMI v1.1 and making some architectural changes to accommodate future growth.

Many users of CMMI v1.1 wanted to extend the model's systematic approach to process improvement to include acquisition and service delivery. CMMI v1.1, however, did not provide the detailed interpretation of process areas, goals, and practices that is needed to address those specific applications. Several reports were written to provide interpretations (Chrissis et al. 2003; Gallagher and Shrum 2004; Herndon et al. 2003), but users still struggled with model details, especially for process areas that fit poorly in the target environment. CMMI v1.2 revises the framework architecture to extend its applicability to new environments. The framework now provides three constellations:

- CMMI for Development (CMMI-DEV)
- CMMI for Services (CMMI-SVC)
- CMMI for Acquisition (CMMI-ACQ)

A *constellation* is defined as "a grouping of model components that are unique to a specific use but also contain a core set of process areas that will not change from constellation to constellation." The constellation concept sets the stage for potential future CMMI expansion. At the time this book was written, CMMI-SVC was still under development.

The CMMI framework consists of the model, training materials, and appraisal methods as well as the rules and methods for generating models, training materials, and appraisal methods. The relationships among those components are shown in Figure 3.1. A constellation consists of a foundation (containing core components), shared material, and specific material. The constellation and its associated training material and appraisal method form a *suite*. The framework may also be extended through the use of "additions." An *addition* is "a clearly marked model component that contains information of interest to particular users." An addition can be a note, a reference, an example, a Process Area, a Specific Goal, or a Specific Practice. As of this writing, there is only one addition, namely, Integrated Product and Process Development (IPPD).

We will return to the architectural details later in this chapter following the discussion of the CMMI v1.2 model components.

As a framework, CMMI is used to plan, define, implement, deploy, benchmark, and improve processes in an organization. The product suite, which contains the models, appraisal methods, and training used to support process integration and product or service improvement, defines the levels through which organizations evolve as they improve their processes. In addition, it enables the definition of process improvement priorities and goals, based on business objectives and goals, and identification of the activities that an organization has to undertake to effectively

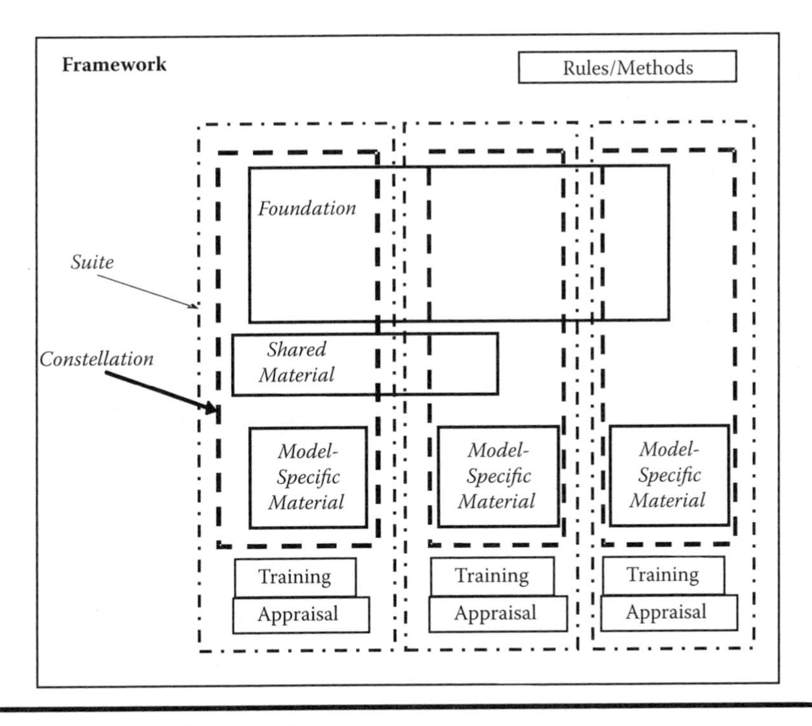

Figure 3.1 CMMI framework components.

and efficiently develop, maintain, and manage software and systems products and services.

It is important to emphasize that CMMI is a model, representing an ideal state that is used as a benchmark; it is not a process. CMMI describes best practices but it doesn't specify how to implement those practices. Organizations have to interpret the model to meet their own applications and develop processes that will be implemented to best satisfy their business objectives. It describes *what* is expected in processes, not *how* to implement processes; definition of the *how* is left to each organization to decide.

Related best practices are aggregated in sets called process areas (PAs). A PA is defined as "a cluster of related best practices that when implemented collectively satisfies a set of goals considered important for making significant improvement in that area."

There are two basic model representations: staged and continuous. They differ in the selection, organization, and presentation of model components but employ the same set of practices. This may be seen as being analogous to different views into a data set in a database. In the continuous representation, process areas are organized in categories (Process Management, Project Management, Engineering, and Support), and in the staged representation they are organized by maturity level (ML 2 through ML 5). A maturity level is a well-defined evolutionary plateau of process improvement across a predefined set of process areas in which all goals in the set are attained. Table 3.1 indicates the distribution of CMMI's 22 PAs across categories and maturity levels, showing that the PAs do not change based on their continuous representation category or their staged representation maturity level.

The continuous representation enables the selection of one or more process areas for improvement. Improvement progress can be measured in terms of capability levels for each selected PA, progressing from an incomplete to an optimizing level

Table 3.1 CMMI v1.2 Process Areas

	Category				
Process Area	*Process Management*	*Project Management*	*Engineering*	*Support*	*Maturity Level*
Causal Analysis and Resolution (CAR)				x	ML 5
Configuration Management (CM)				x	ML 2
Decision Analysis and Resolution (DAR)				x	ML 3
Integrated Project Management (IPM)		x			ML 3
Measurement and Analysis (MA)				x	ML 2
Organizational Innovation and Deployment (OID)	x				ML 5
Organizational Process Definition (OPD)	x				ML 3
Organizational Process Focus (OPF)	x				ML 3
Organizational Process Performance (OPP)	x				ML 4
Organizational Training (OT)	x				ML 3
Product Integration (PI)			x		ML 3
Project Monitoring and Control (PMC)		x			ML 2
Project Planning (PP)		x			ML 2
Process and Product Quality Assurance (PPQA)				x	ML 2
Quantitative Project Management (QPM)		x			ML 4
Requirements Definition (RD)			x		ML 3
Requirements Management (REQM)			x		ML 2
Risk Management (RSKM)		x			ML 3
Supplier Agreement Management (SAM)		x			ML 2
Technical Solution (TS)			x		ML 3
Validation (VAL)			x		ML 3
Verification (VER)			x		ML 3

through satisfaction of a defined set of practices. The order and selection of processes to be improved depend on the business goals and objectives of the organization.

The staged representation enables an organization to select a predefined set of PAs for improvement and uses a maturity level to characterize process improvement. Processes then follow five levels of maturity from initial to optimizing. Capability and maturity levels are shown in Table 3.2 and a comparison of the staged and continuous representations is shown in Table 3.3.

The notion of representation is more important from the appraisal point of view than from the process improvement point of view. During process improvement activities, most organizations select and prioritize PAs based on their process improvement objectives. For example, even if an organization prefers the staged representation, it will not improve all seven maturity level 2 PAs at once. Instead, it will select those that will provide the most immediate value for achieving the process improvement goal. Appraisals are discussed in Chapter 7.

Table 3.2 Capability and Maturity Levels

Level	*Capability Levels (Continuous Representation)*	*Maturity Levels (Staged Representation)*
0	Incomplete	(Not applicable)
1	Performed	Initial
2	Managed	Managed
3	Defined	Defined
4	Quantitatively managed	Quantitatively managed
5	Optimizing	Optimizing

Table 3.3 Comparison of Staged and Continuous Representations

Staged	*Continuous*
PAs are organized by maturity levels.	PAs are organized by categories.
Process improvement is measured by maturity levels.	Process improvement is measured by capability levels.
A maturity level represents a degree of process improvement across a predefined set of PAs.	A capability level is the degree of process improvement achieved within an individual PA. An organization may choose to improve any PA or set of PAs and improve them at different rates.
A maturity level rating, as a single number, may be used to communicate achievement of process improvement goal.	Capability level ratings are used to denote achievements for each PA and may be used for comparing an organization's process area profile to a target profile.

When a process is ingrained in the way an organization does business, we say that the process is institutionalized. This means that the project staff uses defined project processes throughout the product or service life cycle even during the times of stress, such as when the release date is approaching and testing activities have not yet been completed.

As an organization advances in its process improvement activities, its processes mature. For example, at the *Initial* maturity level or the *Performed* capability level, practices may be performed but implementation may be undisciplined. For example, the organization may lack policies, the project may not be planned, progress may not be tracked, staff may not be trained (or their training may be too reliant on on-the-job training and casual mentoring), processes and work products may not be objectively reviewed, and work products may not be kept under control as they are developed. The next maturity or capability level: the *Managed* level, is characterized by the extent to which processes are planned and tracked and corrective actions are taken based on defined criteria. By advancing to the next higher level, the *Defined* level, organizations can start relying on standard process descriptions that are tailored to meet the needs of individual projects. In addition, at the Defined level organizations establish repositories for their work products and measurements so that the collected information may be shared across the organization. At the *Quantitatively Managed* level, an organization starts to understand process behavior and measure its performance against organizational quality and process performance objectives. Organizations and their projects may then be able to statistically control those processes and thus achieve greater predictability of process performance. Since all processes exhibit some normal variation in performance, the next objective in process improvement, at the *Optimizing* level, is to reduce those common causes of process variation, thus tightening the natural limits between which the process operates and preventing occurrence of undesirable process deviations. Performance at the Optimizing level allows an organization to quantitatively understand the effect of process improvement changes.

The description of each PA in the model follows the same basic pattern shown in Figure 3.2. Each PA description starts with a statement, typically one sentence, indicating the purpose of the process area. The next section provides introductory notes that describe the PA and supply context for the descriptions of specific goals and practices. The introductory notes typically summarize the processes that form the basis for the PA. This section contains a wealth of information and should be read carefully. The Related Process Area section lists the most important PA interfaces. Additional references to related PAs may be shown in the PA text to provide information needed to clarify a specific practice or a subpractice.

The next section lists the Specific Practices (SP) grouped by Specific Goal (SG). Each specific goal contains a set of specific practices that, when implemented together, describe a major portion of the PA. Most SPs have subpractices that provide implementation guidance. Implementation of SPs and subpractices will generate work products. Examples of those work products are listed in the section titled

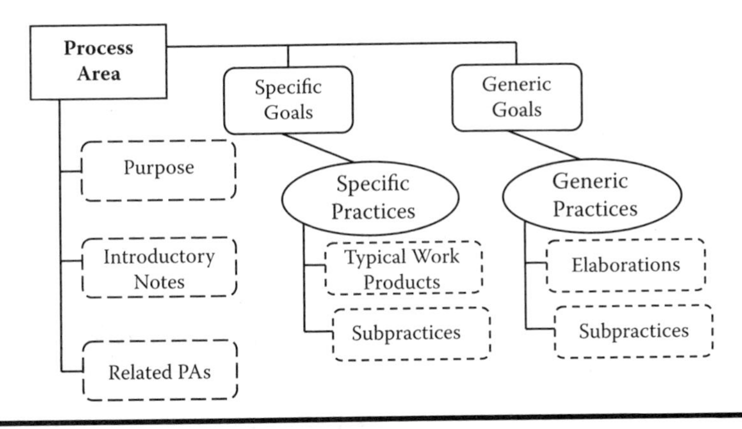

Figure 3.2 Process area structure.

Typical Work Products. Additional information for particular disciplines, such as systems engineering, software engineering, or hardware engineering, is provided through specially marked "discipline amplification" text. These amplifications provide examples specific to that discipline, hints for implementation, or list work product nomenclature variations. Many SPs also include examples to clarify concepts.

Generic Goals (GG) describe how well the SGs in a PA are implemented and institutionalized. Unlike Specific Goals, Generic Goals are cumulative, which means that GG 2 subsumes GG 1, GG 3 subsumes GG 2 (which already subsumed GG 1), and so on. Thus, GGs cannot be skipped. Generic Practices (GP) are associated with the generic goals and describe the detailed activities that are needed for institutionalization. They are generic because they present a uniform approach to institutionalization and appear in each PA with identical numbering and wording. GGs and GPs are described in detail in the model document, prior to the presentation of PA details. Additional elaboration may be provided in each PA to describe how a GP may be interpreted and applied to that PA.

The numbering scheme used throughout the model is quite simple. SGs and GGs are numbered sequentially. The SPs and GPs associated with those goals have the goal number followed by the sequential number for each practice. For example, Generic Goal 2, GG 2, has ten generic practices numbered GP 2.1 through GP 2.10. Specific Practices follow a similar numbering scheme, where, for example, the SPs for SG 3 are numbered SP 3.1, SP 3.2, and so on. GGs and their associated GPs are listed in Table 3.4.

Among all the model elements, only the SGs and GGs are considered to be required components for obtaining necessary capability or maturity information during appraisals. SPs and GPs are expected components, meaning that they can be used to satisfy the SG and GG requirements, but alternatives may be acceptable. All other model elements, such as examples, elaborations, and typical work products are informative. These distinctions are of greatest interest during appraisals,

Table 3.4 Generic Goals and Practices

Generic Goal	*Generic Practice*
GG 1 Achieve Specific Goals	GP 1.1 Perform Base Practices
GG 2 Institutionalize a Managed Process	GP 2.1 Establish an Organizational Policy
	GP 2.2 Plan the Process
	GP 2.3 Provide Resources
	GP 2.4 Assign Responsibility
	GP 2.5 Train People
	GP 2.6 Manage Configurations
	GP 2.7 Identify and Involve Relevant Stakeholders
	GP 2.8 Monitor and Control the Process
	GP 2.9 Objectively Evaluate Adherence
	GP 2.10 Review Status with Higher Level Management
GG 3 Institutionalize a Defined Process	GP 3.1 Establish a Defined Process
	GP 3.2 Collect Improvement Information
GG 4 Institutionalize a Quantitatively Managed Process	GP 4.1 Establish Quantitative Objectives for the Process
	GP 4.2 Stabilize Subprocess Performance
GG 5 Institutionalize an Optimizing Process	GP 5.1 Ensure Continuous Process Improvement
	GP 5.2 Correct Root Causes of Problems

although they may help guide process improvement efforts as well. We will revisit this issue in detail in chapter 7, "Appraisals."

CMMI distinguishes between basic PAs and the advanced PAs, which require more sophisticated constructs. Advanced PAs build on the concepts described in the basic PAs, as shown in Table 3.5.

An important concept to be considered is the relationship between GPs and PAs. Consider the PAs shown in Table 3.1 and the GPs shown in Table 3.4. Based on their titles alone, one can see that there are PAs that are closely related to certain GPs. For example, the Configuration Management (CM) PA will, if properly implemented, satisfy GP 2.6 Manage Configurations. Similarly, the Process and Product Quality Assurance (PPQA) PA can satisfy GP 2.9 Objectively Evaluate Adherence. The detailed relationships among the GPs supporting GG 2 and GG 3 and PAs are shown in Figure 3.3, using the IDEF0 notation. The graphical language of IDEF0 contains two basic constructs: boxes, representing activities

Table 3.5 Basic and Advanced Process Areas

Category	*Basic*	*Advanced*
Project Management	• Project Planning • Project Monitoring and Control • Supplier Agreement Management	• Integrated Project Management • Risk Management • Quantitative Project Management
Support	• Configuration Management • Process and Product Quality Assurance • Measurement and Analysis	• Decision Analysis and Resolution • Causal Analysis and Resolution
Process Management	• Organization Process Focus • Organizational Process Definition • Organizational Training	• Organization Process Performance • Organizational Innovation and Deployment
Engineering	• Requirements Management • Requirements Development • Technical Solution • Product Integration • Verification • Validation	• None

or functions; and arrows, representing inputs, controls, outputs, and mechanisms. The diagram structure supports the presentation of increasing levels of detail for each activity box. IDEF0 diagrams are built by connecting boxes and arrows representing inputs (connected to the left side of the box), controls (connected to the top), mechanisms (connected to the bottom), and outputs (connected at the right side of the box). A control arrow describes something that guides, determines, or constrains the function (e.g., a plan), whereas the mechanism arrow represents the resources that enable the function to operate (e.g., tools). Boxes and arrow segments are combined in various ways to form diagrams. Each box can be further decomposed into subfunctions with their own set of inputs, controls, mechanisms, and outputs. Such decomposition starts with a context diagram that represents the highest level in the hierarchy and may result in hierarchically arranged layers of IDEF0 diagrams.

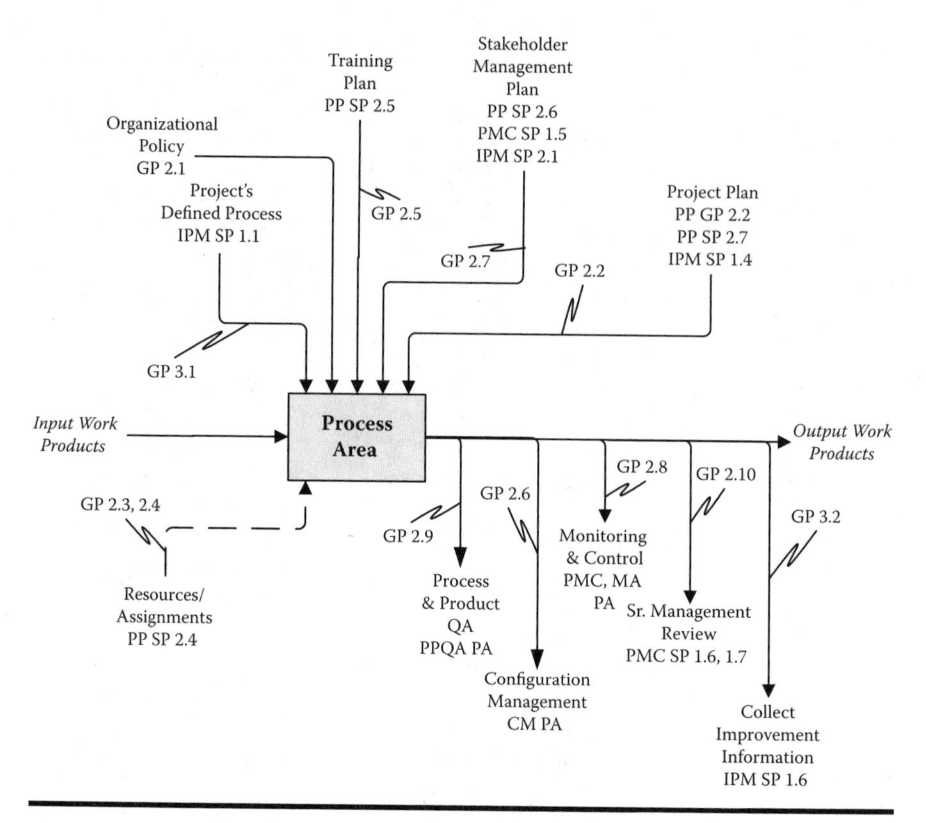

Figure 3.3 PA–GP relationships.

Process Areas

In the following discussion, we group PAs by their affinity, similar to the categories used in the continuous representation. Rather than strictly following the continuous representation categories, we defer the discussion of higher maturity PAs until after the presentation of other PAs. In our brief descriptions of the PAs we will concentrate on the underlying principles and concepts rather than repeating the SP and SG text. We will address major features of the selected PAs as a group and discuss the characteristics and expected work products of each PA. Discussions of product development also include service delivery. The detail presented is sufficient for the reader to understand the characteristics of each PA, compare PAs to the corresponding sections of other frameworks, and understand similarities and differences among the frameworks. For more details on each PA, the reader is referred to the CMMI report (CMMI 2006) or CMMI book (Chrissis et al. 2006).

It is important to remember that CMMI is a model for process improvement. The model is not a process nor is it a life cycle model. The processes discussed in each process area:

- Are intended to be used throughout the life cycle
- Are not tied to any life cycle model or development stage
- Can be applied to both product development and service delivery

Figure 3.4 shows a top-level view of process categories and their relationships, based on IDEF0 notation. We'll start with the Engineering PAs and then address the Project Management PAs, which are used to manage the development, followed by the Support PAs that support all other processes. Next, we'll discuss the Process Management PAs, which provide the organizational infrastructure necessary for PA implementation and institutionalization. We will then address the PAs at higher maturity levels and, finally, we'll address the Integrated Product and Process Development addition.

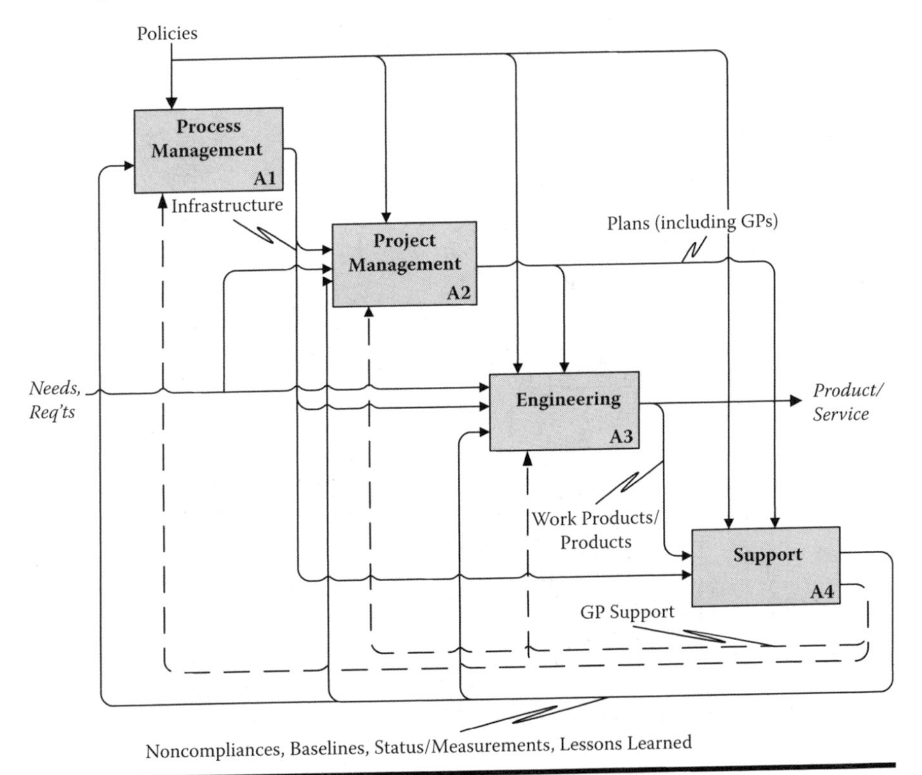

Figure 3.4 Top-level view of CMMI.

Engineering

This group contains the closely related Requirements Development (RD), Requirements Management (REQM), Technical Solution (TS), Product Integration (PI), Verification (VER), and Validation (VAL) PAs. These PAs are shown in Figure 3.5, which is a simplified decomposition of the Engineering process, A3, shown in Figure 3.4. We have added a customer interface, which was not shown at the higher level. The data flows labeled "GP Support" contain all of the relevant GPs listed in Table 3.4.

These process areas are frequently, albeit mistakenly, thought of as constituting a "waterfall" life cycle model. To the contrary, CMMI specifically states that "requirements are identified and refined throughout the phases of the product life cycle" and, when PAs are implemented, "their associated processes may be closely tied and performed concurrently."

Requirements are identified and documented in the RD PA and changes to those requirements are controlled through the REQM PA. When requirements are sufficiently defined and stable enough to act upon, the top-level product architecture is developed in the TS PA. Requirements analysis and top-level design may continue iteratively until the design architecture is mature enough to proceed with

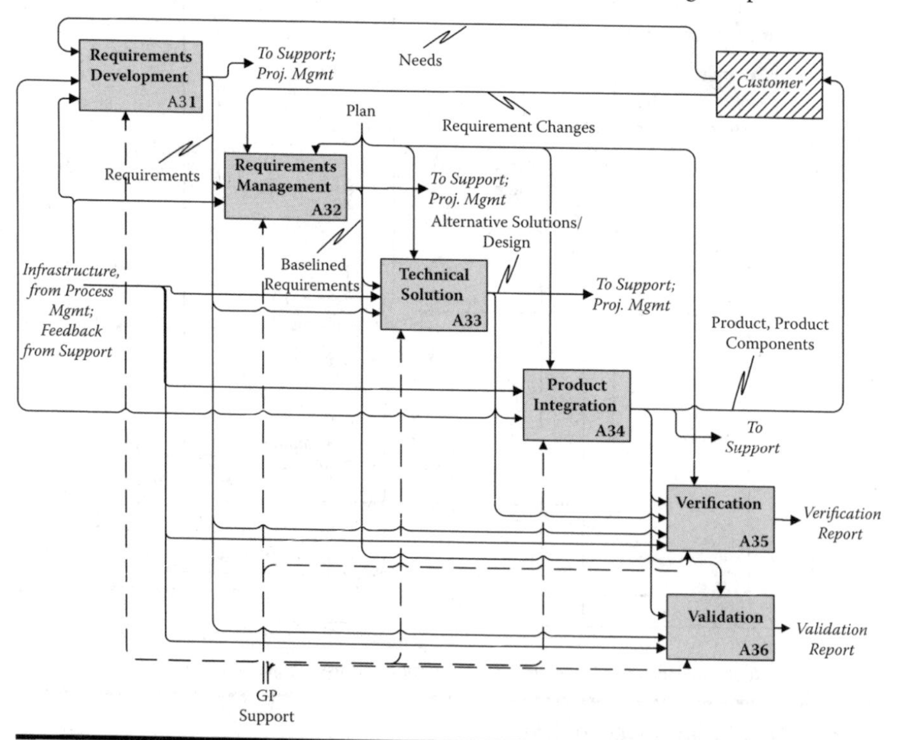

Figure 3.5 Engineering PAs.

detailed component design and implementation. When product components are implemented, they are integrated using processes described in the PI PA. The VER and VAL PAs are implemented throughout the life cycle.

Requirements Development (RD)

Characteristics and expectations: When documenting requirements, it is not sufficient to merely gather them from customers and users. Rather, requirements must be proactively elicited by understanding stakeholders' needs and intentions, analyzed by considering operating conditions and possible constraints, decomposed so that they can be allocated to components and functions, and finally validated to ensure that they will successfully satisfy stakeholder needs in the intended operational environment. At that point, detailed design and implementation, including manufacturing, may start. This, however, does not mean that all requirements must be fully defined before design may start. Different techniques and methods may be used throughout engineering development to enable systematic processing of requirements development, product design, implementation, and testing in an iterative manner. The RD process is closely related to the design, implementation, verification, validation, and configuration management processes.

Expected work products: Requirements definitions (including product, product component, derived, and interface requirements), design constraints, operational concepts and scenarios, and analyses of requirements. Product components are defined as lower-level components of the product. One organization's product component may be another organization's product.

Requirements Management (REQM)

Characteristics and expectations: All documented product and product component requirements must be managed and controlled. This does not mean that the requirements are "frozen," but if and when they change, the project should control acceptance of those changes and account for the impact of the change. The impact potentially includes changes to project plans and downstream work products such as design, implementation, test plans, and procedures. In order to know which changes to implement and which work products will be affected, requirements and work products need to be related through a concept called traceability. The ability to trace requirements in both directions, for example, from requirement specifications to test procedures and from test procedures to requirement statements, is referred to as bidirectional traceability.

Expected work products: Requirements changes and their status, impact of changes, requirements traceability, detected inconsistencies and their resolution.

Technical Solution (TS)

Characteristics and expectations: As noted earlier, the TS PA should be considered together with the other engineering PAs and they should be implemented iteratively and recursively. Recursively means that the processes may be implemented at all levels of product decomposition, from the high-level architecture to the lowest levels of implementation. The TS PA consists of two major subprocesses: design (SG 1 and SG 2) and implementation (SG 3) that can be equally applied to products, product components, services, and product-related life cycle processes (e.g., a manufacturing process required to implement specific portion of the design). The design processes deal with:

- Selecting and evaluating solutions to satisfy the requirements allocated to a specific portion of the product architecture
- Making decisions about selecting, purchasing, or reusing certain product components
- Evolving the concept of operations
- Developing a preliminary design with clearly articulated product architecture
- Developing a detailed design of product components

The implementation process deals with converting the design into a product or product component. The implementation process involves activities such as allocation, refinement, and verification of each product component from a high level in the product hierarchy to the actual manufacture of the product component, or development and test of software code. The TS PA includes practices dedicated to the development of the technical documents needed to implement the design (or procure a product meeting the design specifications) and the product support documentation required to install, operate, and maintain the product.

Expected work products: Alternative solutions, selected solution and the rationale for its selection; operational concepts and scenarios; product architecture; make-or-buy and reuse analyses; product component design; design data package necessary for performing implementation; implemented design; and documentation needed for installing, operating, and maintaining the product or service.

Product Integration (PI)

Characteristics and expectations: This PA deals with assembly of the implemented product components into completed products that will be packaged and delivered to customers. Throughout the integration process, the interfaces among the components to be assembled into products must be managed. When changes in the upstream product components occur, the interfaces among the downstream components may be impacted and will have to be addressed. Organizations will typically establish environments for product integration, which may range from a single laptop computer up to an elaborate integration facility using simulators or emulators to test integrated product components. Upon completion of integration and testing, products are packaged and delivered to the customer.

Expected work products: Integration strategy, integration facility and/or environment, updated interface documents, assembled product, packaged product, and delivery documentation.

Verification (VER)

Characteristics and expectations: Verification occurs throughout the product development life cycle, from requirements analysis to completed product or service. The verification process includes:

- Selection of the work products to be verified
- Identification of requirements to be satisfied by the selected work product
- Selection of verification methods, including reviews, testing, demonstration, prototyping, simulation, and analysis
- Establishment of the verification environment
- Creation of verification procedures, including acceptance criteria
- Execution of the verification procedures and analysis of the obtained results

Verification is similar to validation but the two processes are used differently during product development. Verification is used to establish that the work product faithfully implements its requirements, whereas validation is used to ensure that the product satisfies its intended use. In CMMI, the management and disposition of verification findings is allocated to the Project Monitoring and Control (PMC) PA (SG 2), although most organizations will perform those functions together with the verification process itself. The VER PA includes the peer review process (SG 2), which is a very effective verification method used to detect and eliminate both product and process defects early in the product life cycle.

Expected work products: Selection of work products to be verified, methods to be used for verification, verification environment, verification procedures including acceptance criteria, verification results and reports, defects identified, peer review information (such as schedule, checklists, work-product size, results, time spent in peer review preparation and execution, and number of reviewers).

Validation (VAL)

Characteristics and expectations: As previously noted, the validation process is very similar to the verification process. Validation is usually applied to products in their intended operational environment and often includes direct customer involvement. In many cases both verification and validation are executed concurrently and may use the same facilities as product integration. Nonfunctional requirements specified at the onset of a project will, in most cases, contain validation requirements, including requirements for integration, verification, and validation facilities. The validation process includes:

- Selection of products to be validated
- Selection of validation methods, similar to those used for verification
- Development of validation procedures and acceptance criteria
- Establishment of the validation environment
- Execution of validation procedures and collection of validation information

Expected work products: Identification of products to be validated, validation methods, validation environment, validation results, and reports.

Summary and Comments

The Engineering PAs are, in general, executed iteratively. The processes associated with each PA are mapped onto the project's selected life cycle model and may be implemented multiple times.

Project Management

Project management processes are used to manage other project processes. The basic project management processes are Project Planning (PP), Project Monitoring and Control (PMC), and Supplier Agreement Management (SAM). The advanced processes are Integrated Project Management (IPM), Risk Management (RSKM), and Quantitative Project Management (QPM). The advanced project management processes are built on the foundation provided by the basic processes. We will address QPM with the other higher maturity PAs. Adding more rigor and better management techniques, such as sharing information among projects in the organization, managing stakeholders, and performing rigorous risk management to the basic management processes helps the organization achieve more effective management processes. It may be advantageous to consider the basic and advanced PAs together when planning and implementing process improvements. The project management processes and their relationships discussed in this section are shown in Figure 3.6. We will refer to this section when appropriate to provide the link between the project management PAs and the higher maturity PAs. Note that GP 3.1 and GP 3.2 have special significance when moving from CL/ML 2 to CL/ML 3. We will discuss this later in the chapter.

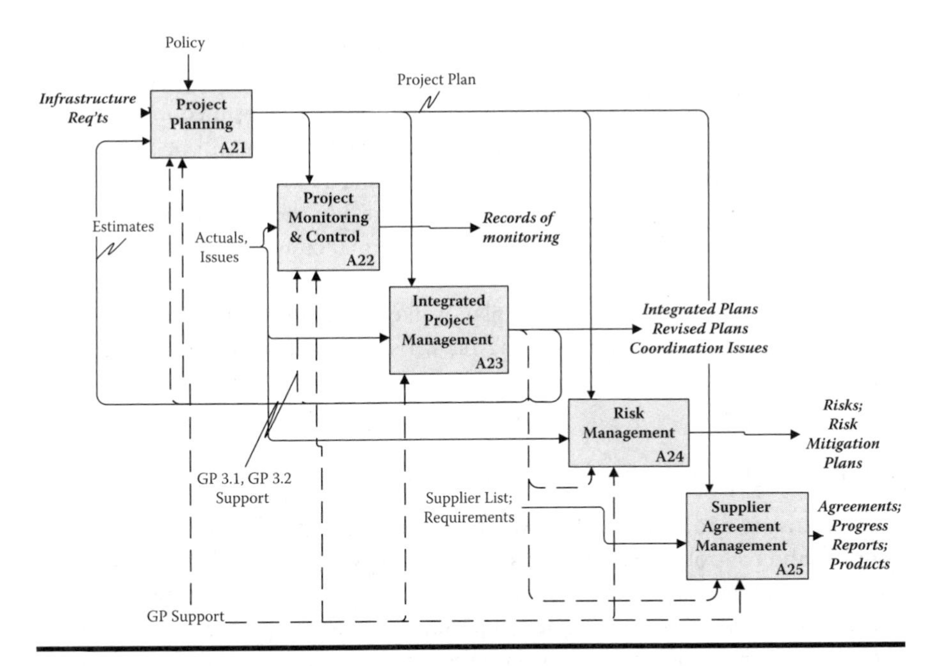

Figure 3.6 Project Management PAs.

Project Planning (PP)

Characteristics and expectations: The project planning process covers several basic project management steps: develop estimates, develop the plan, and get commitment to the plan from the various stakeholders. The essential concepts of the PP PA include:

- Defining the scope of work, based on an understanding of the requirements
- Determining the work products that will be produced
- Developing estimates of work product attributes such as size (typical size attributes are lines of code, number of function points, number of requirements, and number of gates in a logic diagram)
- Deriving effort estimates
- Identifying resource and budget requirements
- Specifying the life cycle, technical approach, and schedule
- Identifying risks
- Planning for data management, training, and stakeholder involvement
- Summarizing project plans in a written plan

The written plan may be a single document or a set of several documents and subordinate plans. The selected planning style will depend upon several factors, such as the size of the project, criticality, precedence, and organizational and customer requirements. It is important to emphasize that stakeholders, both internal and external, have to commit to the plan. As an organization matures, its planning processes will also mature and estimates will be based on historical data stored in the organization's measurement repository. In addition, project processes will be tailored versions of the organizational standards, and its stakeholders will be effectively managed according this defined process. As shown in Figure 3.6, the Project Planning PA is closely related to the PMC PA, which tracks performance against the plans, and the IPM PA, which provides organizational information.

Expected work products: Project plan, stakeholder involvement plan, resource plan, data management plan, training plan, risk management plan, budget, and schedule.

Project Monitoring and Control (PMC)

Characteristics and expectations: The plans developed in the PP PA are used to monitor and control project activities. Monitoring is done through progress reviews and formal milestone reviews. Progress in achieving the project's goals and objectives is measured by comparing actual performance to the plan. If those comparisons show significant deviation from the plan, corrective actions are taken and managed to closure. The definition of "significant deviation" is a function of the project's and organization's process capability or maturity. At CL or ML 2 the significant deviation criteria are generally defined by the project during project planning. As organizations and projects mature to CL/ML 3, this deviation may be specified as a percentage band around the planned value. At CL/ML 4 and 5, the thresholds associated with the band are known quantitatively and depend on actual process performance. Such limits are called the "voice of the process."

Expected work products: Records of project performance (such as measurements of effort, budget, schedule, and staffing), monitoring of risk, stakeholder involvement, training records.

Integrated Project Management (IPM)

Characteristics and expectations: IPM is an advanced project management PA that extends the practices identified in the PP and PMC PAs. The IPM PA has two SGs that address the use of the project's defined process and enhances the coordination and collaboration with stakeholders initially established in the PP and PMC PAs.

One of the most important IPM practices is the tailoring of the organization's set of standard processes to fit the project's specific requirements. Tailoring is based on the organization's tailoring criteria and guidelines, which take a project's size, complexity, safety, security, and impact on the bottom line into account. The result of tailoring is an integrated, defined project process that uses the organization's library and database for establishing its estimates, developing its process documentation, and integrating its plans into an effective whole. The integrated plan is then used to manage the project, manage stakeholder involvement, and resolve coordination issues. In order to close the process improvement loop, IPM processes support the collection of a project's work products, measures, and lessons learned and require their submittal to the organization's library and database so other projects may benefit from organizational experiences.

The IPPD addition provides for the establishment of a shared vision for the project and the implementation of an empowered, integrated team structure. An integrated team draws resources from all involved areas to ensure that the overall needs of the business, and not only the needs of the project, are addressed. When using the Integrated Product and Process Development (IPPD) addition, a third SG is used to support the application of IPPD principles, such as design of downstream processes during product design, focus on the customer's needs during product and process development, timely and appropriate collaboration of all relevant stakeholders, continuous and proactive identification and management of risk, and focus on measurement and process improvement. The IPPD principles that needed specific implementation have been added to the IPM and OPD (Organizational Process Definition) PAs. They are:

- Effective use of cross-functional or multidisciplinary development teams
- Leadership commitment
- Appropriate allocation and delegation of decision making
- An organizational infrastructure that supports and promotes IPPD concepts

As shown in Figure 3.6, the IPM PA and the related generic practices, GP 3.1 and GP 3.2, play a central role in the feedback loop needed for continuous process improvement.

Expected work products: Project's defined process, integrated project plans, collected measures, work products, lessons learned, documented project and stakeholder issues, critical dependencies and their status.

Risk Management (RSKM)

Characteristics and expectations: Risks are initially identified and monitored in the PP and PMC PAs, respectively. RSKM, an advanced project management PA, extends those basic and largely reactive processes to implement continuous risk management. Under continuous risk management, risks are proactively identified early in the project's life cycle, and are continuously updated, categorized, evaluated, prioritized, and mitigated as needed. This PA establishes the basis for systematic risk management by requiring the identification of the risk sources typically encountered by the organization. This is often implemented through the use of a risk taxonomy. A consistent approach to risk management is supported by defining the parameters to be used, such as probability and impact, and the actions to be taken at different risk exposure levels, such as development of mitigation or contingency plans. Finally, an overall strategy should be defined to identify scope, responsibility, tools, and timing for risk management activities.

With this foundation in place, a project can implement the risk management strategy, continuously identifying, evaluating, and prioritizing risks. As appropriate, risk mitigation plans and contingency plans are then developed and implemented.

Expected work products: Risk sources and categories, prioritization criteria, thresholds for each risk category, risk management strategy, documented risks with calculated exposure, prioritization, mitigation plans, contingency plans, status.

Supplier Agreement Management (SAM)

Characteristics and expectations: The acquisition of products and services that will be included in, or will support, delivered products or services has a significant impact on the success or failure of a project. This PA deals with products or services created outside the project that will be delivered to the project's customer and for which there is a formal agreement. A formal agreement may be any legal agreement ranging from a memorandum of understanding to a contract. This PA does not address situations where a supplier is used to augment the workforce of a given project. The processes covered by this PA consist of selecting a supplier, establishing the supplier agreement, executing this agreement, and then accepting and transitioning the product. In some organizations, particularly in large organizations, processes described in this PA are documented and implemented by other organizational entities, such as contracts or purchasing departments, and may follow specific regulatory statutes. In those cases, the SAM process description will describe processes implemented by the project itself and identify the interfaces with processes from the other organizational units.

Expected work products: Lists of candidate suppliers, evaluation criteria, rationale for selection, agreements, trade studies, supplier performance reports, action items from reviews, acceptance test results, transition plans.

Summary and Comments

As an organization matures its processes, efforts to achieve better process effectiveness and efficiency are enabled by the IPM process area. Tailoring a common set of processes to meet a project's needs provides the necessary means for better staff training, the ability to move staff from project to project without extensive training, and the capability to react quickly to changes in customer requirements.

Support

Support PAs serve all processes. They are used throughout the project life cycle under the responsibility of the project manager, although they may be managed by separate organizations.

Figure 3.7 shows both basic and advanced Support PAs and their interactions. The Causal Analysis and Resolution (CAR) process area will be addressed in the "Higher Maturity Level Process Concepts" section of this chapter. Figure 3.7 shows a simplified view of the interactions among the support PAs and all other PAs. The plans, shown as flows entering each process, normally address many of the GPs shown in Table 3.4.

Configuration Management (CM)

Characteristics and expectations: The CM PA addresses four major processes: configuration identification, change control, configuration status accounting, and configuration audits. Most projects use automated tools to support some or all of those processes. Unfortunately, there is still confusion about which activities constitute

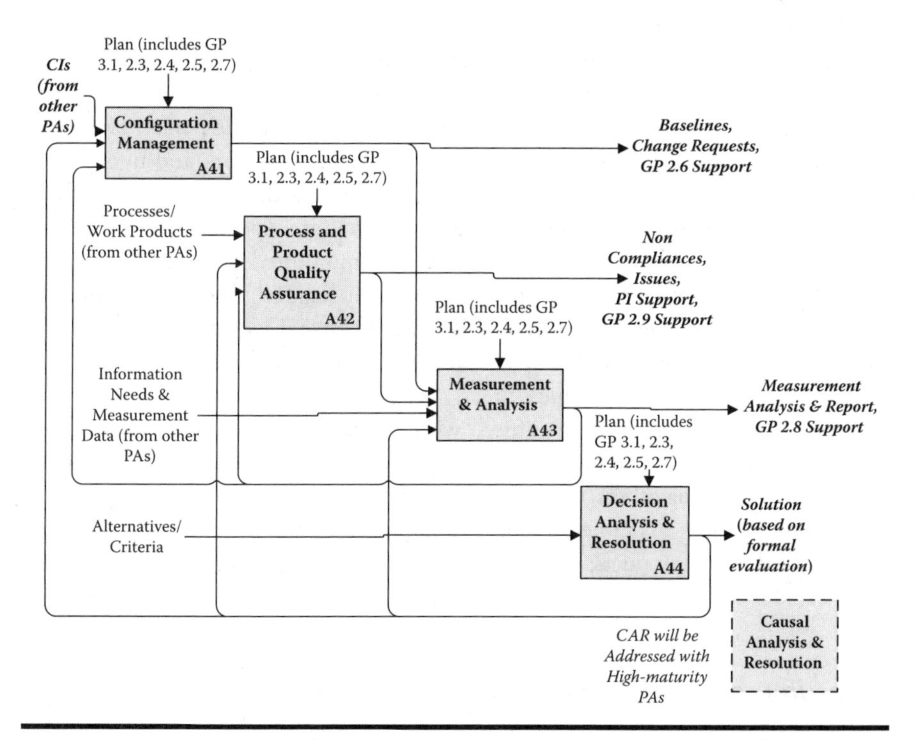

Figure 3.7 Support PAs.

configuration management. The selection of work products that will be under configuration management versus less formal control is important in enabling a project to balance control and efficiency of work product generation, integration, testing, and delivery. Central to this PA is the establishment of a set of baselines, the control of work products associated with that baseline, and the movement of work products from one baseline to another. Change control supports some aspects of CM, but not all. Change control alone may be effective in controlling intermediate work products under development but may not be sufficient for establishing and maintaining the integrity of delivered products or deciding if the product is complete and ready for delivery.

Expected work products: Configuration item identification, configuration management system, baselines, change requests, change histories, configuration status accounting reports, configuration audit reports.

Process and Product Quality Assurance (PPQA)

Characteristics and expectations: PPQA provides for the objective evaluation of both processes and work products to ensure their quality. One way to implement an objective evaluation capability is by establishing an independent group, usually external to the projects, to perform PPQA tasks. However, in organizations with a history of successful process improvement, where everybody participates in the quality assurance process, it may be sufficient to ensure that the person evaluating the processes and work products does not evaluate his own work and uses process descriptions, policies, procedures, checklists, and standards to ensure their objectivity. Any discrepancies identified during evaluations are reported to those that are immediately impacted and corrective actions are monitored to closure. Typically, the QA processes have an escalation clause that may be used if noncompliances are not satisfactorily resolved. The QA staff is most effective if its work results not only in identifying noncompliances but also in preventing them from happening again. In that way, the QA staff becomes directly involved in process improvement.

Quality assurance is often confused with verification and validation processes. The role of the quality assurance, however, is to ensure that the verification and validation processes, among others, were performed as described and that work products are developed according to defined standards and procedures.

Expected work products: Evaluation reports, noncompliance reports, corrective actions and their status, quality trends, and process improvement suggestions.

Measurement and Analysis (MA)

Characteristics and expectations: Everybody agrees that measurement of both processes and products is necessary but establishing an efficient and effective measurement capability is difficult. The most fundamental concept described in this

PA is that measurements at both the project and organization levels are based on management information needs. Identification of information needs starts with the question: What information does management need to have to assess a project's progress, process effectiveness, product quality, and readiness for product release to the customer? This PA requires:

- An understanding of project or business objectives
- Identification of measurement needs related to those objectives
- Specification of measures, collection methods, storage plans, analysis techniques, and reporting plans
- Collection, storage, analysis, and reporting of measurements in accordance with plans

By using techniques such as Goal–Question–Metric (GQM; Basili 1992) appropriate measurements to satisfy the stated management objectives and needs can be developed.

Expected work products: Business objectives, measurement objectives (linked to the business objectives and information needs), measures (including definitions and specifications for collection, storage, analysis, reporting), collection and analysis tools, measurement data, and measurement reports.

Decision Analysis and Resolution (DAR)

Characteristics and expectations: Some decisions require a formal evaluation process to enable selection of a solution from among identified alternatives. Typically, a project will:

- Decide when such a formal evaluation process is needed (for example, for high-risk issues, make-or-buy decisions, or supplier selection)
- Determine alternatives and criteria for making a decision
- Select an evaluation method
- Perform the evaluation
- Document the recommended solution

Processes in this PA can be invoked at any time in the project life cycle, although the degree of formality may vary from issue to issue. An advantage of using a formal process is that the selection is objective and the selection results and rationale are documented. There are many decision analysis methods such as Analytical Hierarchy Process (AHP), Multi-Attribute Utility Technique (MAUT), decision trees, pairwise comparison, or decision matrices. Some methods may be efficient in certain conditions and fail in others, depending on the number of alternatives and criteria. In general, when the number of criteria and alternatives is large, special techniques—

such as Monte Carlo simulation, linear programming, design of experiments, or quality function deployment—are required. Even then, a suboptimal solution may be chosen. It is wise to limit the number of criteria and alternatives to a relatively small number, for example, ten to twenty.

Expected work products: Guidelines for invoking the DAR process, selection alternatives and criteria, selected solution and rationale.

Summary and Comments

Although interaction among the Support PAs is largely limited to the GPs, the interaction of the Support PAs with other PAs is significant. They are instrumental in the effective implementation of the GPs and are shown as the "GP Support" flows on all figures.

Process Management

PAs in the process management category are organization related and deal with establishing the process improvement infrastructure. We will address process management in this section and in the higher maturity PA section. There is no reason why process management PAs should not be implemented in organizations at any maturity or capability level. In fact, this is desirable in most cases (Garcia and Turner 2006; Mutafelija and Stromberg 2003a). The PAs discussed in this section are essential to bringing a process improvement focus to an organization by providing the guidelines for establishing a process architecture, a process asset repository, and measurement database, and by addressing practices for establishing an organizational training capability.

Organizational Process Focus (OPF)

Characteristics and expectations: In order to develop an effective process improvement program, an organization must establish a focal point for its infrastructure implementation. The organization must establish its process improvement needs and goals, evaluate its current processes' strengths and weaknesses, and, based on those characteristics, develop a plan for addressing weaknesses and achieving its process improvement objectives. Process strengths and weaknesses are usually determined by conducting an appraisal to compare an organization to the expectations of reference frameworks such as CMMI or ISO 9001. Appraisal results are used as input to a process improvement plan, which is then used to implement and manage process improvement activities. Sponsorship and implementation responsibility are usually vested in two organizational functions: a *management steering group* that provides vision, funding, and staffing, and an *engineering process group* that facilitates and implements process improvement activities.

Expected work products: Process needs and objectives, process improvement plan, action items resulting from plan reviews.

Organizational Process Definition (OPD)

Characteristics and expectations: An organization that plans to improve its process performance and achieve higher capability or maturity levels needs an infrastructure that will enable projects to efficiently and effectively implement and execute their processes. This infrastructure provides the definition of a set of standard processes that may then be tailored based on the individual project's need. Typically, organizations establish a process architecture and then develop a collection of subprocesses or process elements reflecting the best practices of the organization to populate that architecture. The process architecture enables process users to visualize process element interfaces and their interactions. In addition, organizations establish a set of life cycle models that are applicable to most of their development and service activities, and collect measurements to quantitatively describe the process elements that populate those life cycles. Organizations experienced in process improvement will have several standard processes and guidelines to aid projects in selecting and tailoring (where permitted) appropriate process elements.

The OPD PA is also used to establish work environment standards. These standards may include, for example, specification of organizational standard products for hardware and software, health and safety requirements, and security procedures. Artifacts that are collected from projects are stored in two repositories: a Process Asset Library (PAL), which is used to store process descriptions, plans, practices, and lessons learned; and a measurement database, which is used to store process and product measurements.

When the IPPD addition is used, a second goal is added to the OPD PA to establish rules and mechanisms to enable integrated team activities. In much the same way as the organizational process assets defined in the OPD PA are used to define the project's processes, the guidelines established here for integrated teams become part of the project's defined processes through the IPM PA.

Expected work products: Organization's set of standard processes, approved life cycle models, tailoring criteria and guidelines, process asset library, and measurement repository.

Organizational Training (OT)

Characteristics and expectations: Organizations need well-educated and skilled workforces to effectively compete. The OT PA addresses an organization's strategic training needs. Training needs are developed by analyzing current skill levels and future business needs. When the gaps between those two states are known, the organization can develop its tactical training plans and the necessary training capability, and then provide the training required to close the gaps. As training is delivered, the organization can collect data about changes in workforce qualifications, training efficiency, and training effectiveness.

Expected work products: Training needs, tactical training plan, training materials and associated artifacts, training records, training effectiveness surveys, training course, and instructor evaluation.

Summary and Comments

Organizational PAs provide the process improvement infrastructure. Although they are associated with maturity levels 3 and higher in the staged representation, they should be implemented as early as possible even though their implementation may be less effective than in higher maturity organizations. For example, establishing a process improvement focal point and running process improvement as a project will benefit a maturity level 2 organization as well as it will a maturity level 3 or higher organization.

Higher Maturity Level Process Concepts

When CMM for Software, one of the source models for CMMI, was published in 1991, there was very little information about the behavior of software projects that were statistically managed. After more than 15 years of experience with the use of quantitative techniques for controlling systems and software projects, there is sufficient evidence that quantitative control of such projects is possible, desirable, and effective. In this section, we will address the higher maturity level process concepts. Statistical and other quantitative methods help us understand process performance and are specifically focused on understanding variation. Projects use those statistical and other quantitative techniques for quantitative management activities, such as planning and controlling their process performance. Organizations collect process performance and product quality data and, based on those data, establish process improvement baselines, create process performance models, evaluate the impact of improvements, and set realistic organizational process performance and product quality goals and objectives.

To implement statistical methods, organizations need data accumulated over time. In the context of CMMI, data collection starts at capability or maturity level 2 and continues forever.

Interaction among the higher maturity level PAs and other processes is shown in Figure 3.8. When a sufficient number of data points are collected, organizations can draw conclusions about process performance and make predictions about achieving process and product objectives. At capability or maturity level 4, organizations

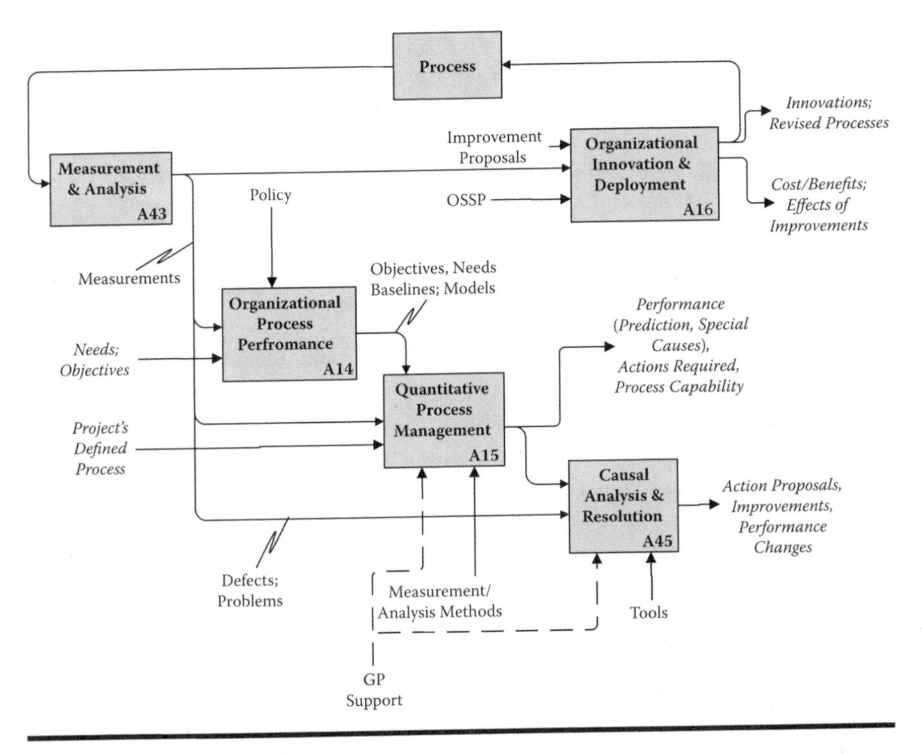

Figure 3.8 High maturity PAs.

establish quantitative process performance and product quality objectives. Projects, using those organizational objectives, select subprocesses that may meet those objectives, monitor their process performance, and address cases where objectives are not achieved. When a process exhibits variations that are normal and expected, we can say that the process is stable and its results are predictable. Based on process performance history we can then predict, with some degree of certainty, that such performance may be achieved in the future. On the other hand, if a process exhibits exceptional variation as a result of causes that are not part of ordinary process performance, we say that such a process is not predictable and that those special causes of variation will have to be addressed, understood, and eliminated.

Organizational Process Performance (OPP)

Characteristics and expectations: When an organization understands the inherent variation in its standard processes, it can establish quality and process performance objectives that projects can use to quantitatively manage their processes. Since it is not possible to manage all subprocesses, the organization will have to select those processes that are the most likely to affect its stated objectives and that can be successfully measured and understood. By collecting and analyzing data to understand process behavior, organizations can establish process performance baselines and models, and use those models to predict the likelihood that projects will achieve their performance objectives and goals.

Expected work products: Organization's quality and process performance objectives, processes and associated measures, process performance baselines and models.

Quantitative Project Management (QPM)

Characteristics and expectations: Based on organizational and project-specific quality and process performance objectives, projects construct their defined process and select the subprocesses that will be quantitatively managed to achieve those objectives. Projects collect measurements, compare the measurements to defined objectives, analyze results using statistical techniques, and determine if the process is capable of achieving its quality and performance objectives. If there are special causes of variation that exceed the process's natural bounds, projects analyze those instances and remove the cause of variation, thus achieving the stability that will enable them to predict process performance and product quality. The observed process behavior is also called the "voice of the process" and the targeted objectives are called the "voice of the customer." The difference between the actual and desired performance determines process capability: a process that is able to satisfy the desired performance is called "capable."

Expected work products: Project's quality and process performance objectives, candidate subprocesses for statistical management, measurements and associated analytical techniques needed for statistical process management, collected measures, calculated natural process bounds, process performance, and process capability.

Organizational Innovation and Deployment (OID)

Characteristics and expectations: This advanced Process Management PA builds on the practices of the OPF PA. It enables the selection and deployment of those process improvements that may enhance an organization's ability to meet quality and process performance objectives, based on a quantitative understanding of the costs and benefits of the proposed improvements. In this PA, improvement proposals and proposed innovations are analyzed, incorporating quantitative knowledge of stable and predictable processes or subprocesses. Selected proposals are piloted to down-select improvements that will be deployed across the organization. After improvement proposals are successfully implemented, the effect of their implementation is measured, and costs and benefits are captured.

Expected work products: Process and technology improvement proposals, candidate innovations, pilot reports, measurements of the effects of process improvement.

Causal Analysis and Resolution (CAR)

Characteristics and expectations: Causal analysis and resolution processes are used to identify and eliminate causes of selected defects and other problems. Because not all defects or problems require such analysis, a cost–benefit analysis should be performed to select the issues that will potentially have the largest impact. Causes of defects may be most effectively eliminated when processes are quantitatively understood and stable. By reducing common causes of process variation, the process natural bounds are reduced, resulting in more accurate predictability and elimination of "chronic waste." Root cause analysis doesn't stop at identifying and eliminating the causes of defects but also seeks to prevent future recurrence. Causal analysis is initially performed at the project level and then, if proven to be effective, is extended to the whole organization.

Expected work products: Defects and problems selected for analysis, action and improvement proposals, causal analysis and resolution records, measurements of performance and performance change.

Summary and Comments

Higher maturity PA implementation presupposes the stability and statistical control of processes. However, CAR may also be implemented at lower maturity levels when addressing problems and defects. Similarly, a project that is not yet ready for quantitative management may also benefit by implementing statistical methods, in the manner of Six Sigma, when analyzing its collected data.

Process Area Interactions

An understanding of PA relationships is important for process improvement. Figure 3.9 shows an example of those relationships, addressing interactions between an arbitrary Engineering PA and the CM PA. The Engineering PA accepts a series of inputs from other Engineering, Support, or Project Management PAs. Both the Engineering PA and the CM PA may be constrained, for example, by policies, plans, and the project's defined process. The CM process and GPs supported by other PAs, such as PMC, PPQA, and MA, provide feedback to the Engineering PA. The Engineering process provides outputs to other PAs and to the CM PA in the form of work products. In this example, the project's defined process, developed using GP 3.1 Establish a Defined Process, is implemented in both the CM PA and the Engineering PA through plans governing their implementation and performance.

The CM PA interacts with all Engineering PAs because it controls the engineering work products to be released, as well as intermediate work products through GP 2.6 Manage Configurations. Similar interaction diagrams can be constructed for all PAs.

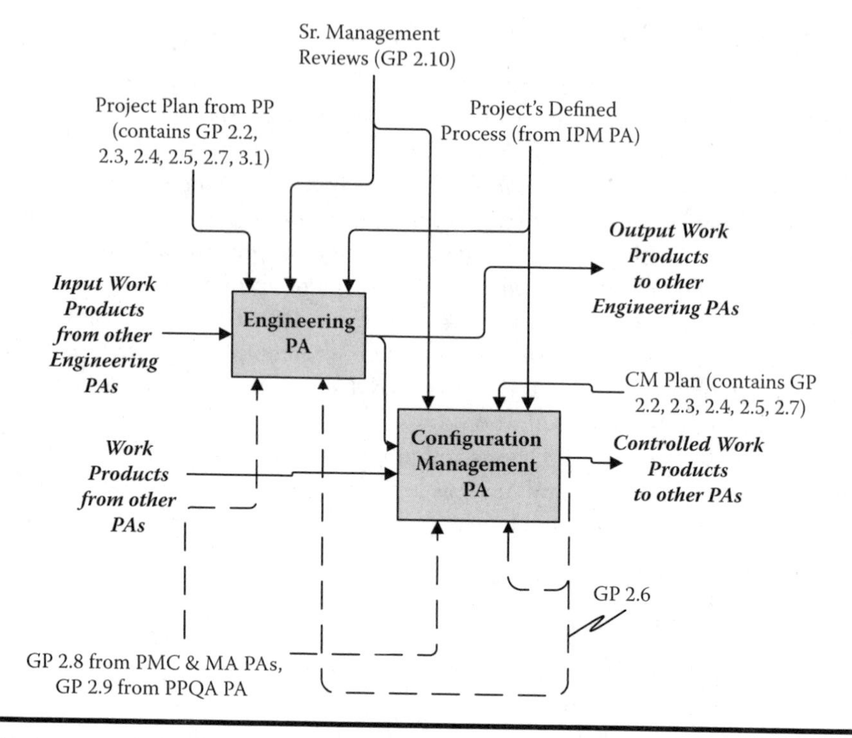

Figure 3.9 An example of PA interaction.

Identification of PA to PA and PA to GP interactions facilitates definition of a process architecture and provides an efficient way for constructing process elements and defining their interfaces. In addition, these relationships also highlight certain constraints when using the continuous representation. For example, implementation of GP 3.1 Establish a Defined Process requires implementation of IPM SP 1.1 "Establish and maintain the project's defined process from project's startup through the life of the project," which, in turn, requires implementation of the OPD PA, which provides a set of organizational standard processes and tailoring guidelines and criteria.

Continuous Representation Considerations

We often hear these questions:

- Which representation should I use?
- Which is best?
- Which representation will help me reach my objectives?
- Can we get a maturity level rating if we use the continuous representation?

These are all valid questions with, unfortunately, no simple answer other than the perennial standby of the process improvement business: "It depends."

In the process improvement literature there are several attempts to clarify the selection process and benefits of each representation (Hefner 2004; Kasse 2002, 2004; Kulpa and Johnson 2003) and we will not repeat them here. Selection of the model representation actually has more to do with appraisal considerations than with process improvement, so let's address the process implications of the selection of the representation first.

As shown earlier in this chapter, there are many strong relationships among PAs and between the PAs and GPs. Those relationships constrain an organization's efficient selection and implementation of processes. For example, efficient implementation of GP 2.6 Manage Configurations would benefit from implementation of the CM PA. Therefore, it seems natural to strive to implement the CM PA early so it may be used to provide the infrastructure for controlling work products across all other processes. The situation is similar with GP 2.9 Objectively Evaluate Adherence, which benefits from implementation of PPQA. We can continue addressing those processes one by one, but, in general, it seems that the concept of institutionalizing maturity level 2 PAs before addressing other processes is not such a bad idea. Do we need to implement all the bells and whistles of the CM or PPQA PAs for them to be useful in process improvement? Of course not, but it certainly helps if there is a plan for those PAs, resources are allocated, assignments are made, training is provided, and implementation is monitored and reviewed.

When we move to higher maturity levels, the problems get more complicated. To achieve CL 3 in any PA we need to implement GG 3 Institutionalize a Defined Process. GG 3 has two associated GPs: GP 3.1 Establish a Defined Process and GP 3.2 Collect Improvement Information. Implementing GP 3.1 requires, at a minimum, the organizational set of standard processes (OSSP) and the tailoring guidelines, which are developed in the OPD PA. Therefore, at least some portions of the OPD PA must be implemented. Implementing GP 3.2 requires a process asset library and a measurement repository, both of which are components of the OPD PA. Implementation of IPM SP 1.6 Contribute to the Organizational Process Assets may also be needed. Clearly, certain sequences of PAs and process implementation are better than others.

The situation gets even more complex when we use the continuous representation and attempt to move a PA to CL 4 or CL 5. To achieve CL 4, GG 4 Institutionalize a Quantitatively Managed Process must be implemented for the selected PA. GG 4 has two associated GPs: GP 4.1 Establish Quantitative Objectives for the Process and GP 4.2 Stabilize Subprocess Performance. If we now revisit the OPP PA we notice that SP 1.3 "Establish Quality and Process–Performance Objectives" and QPM SP 1.1 "Establish the Project's Objective" support implementation of GP 4.1 in all PAs. Similarly, QPM SG 2 Statistically Manage Processes and the entire OPP PA support GP 4.2. A similar situation is encountered when moving to CL 5.

One can use the continuous representation and still claim maturity levels. This is where equivalent staging comes into consideration. To achieve ML 2 using the continuous representation, all PAs that are associated with ML 2 have to achieve CL 2. To reach ML 3, CL 3 must be achieved in all PAs associated with ML 2 and 3. To reach ML 4, the organization will have to achieve CL 3 in all PAs associated with ML 2, 3, and 4. The achievement of ML 5 requires reaching CL 3 in all PAs.

The reason that there is no requirement for achieving GG 4 and GG 5 when using equivalent staging is that GP 4.1 and GP 4.2 are satisfied by OPP and QPM, which are ML 4 PAs in the staged representation. GP 5.1 and GP 5.2 are satisfied by OID and CAR, which are ML 5 PAs in the staged representation.

CMMI v1.2 Constellations

As noted earlier in this chapter, a constellation consists of a foundation, shared material, and specific material, as illustrated in Figure 3.10. To ensure commonality among constellations, the model foundation contains the set of core PAs shown in Table 3.6, all generic goals and practices, selected portions of the front matter, and a core glossary. Each constellation must use those components without change or

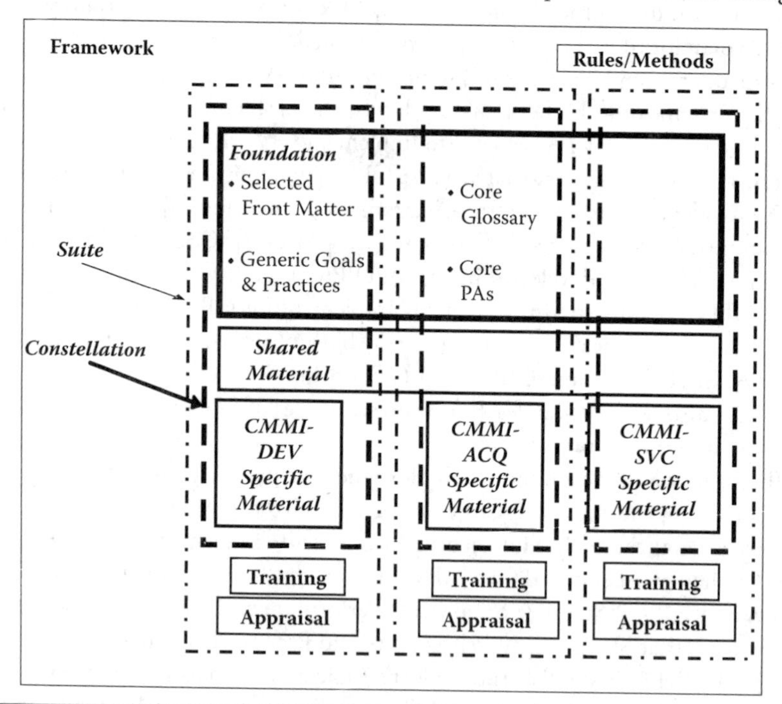

Figure 3.10 CMMI v1.2 architecture.

Table 3.6 CMMI v1.2 Core Process Areas

Causal Analysis and Resolution (CAR)	Organizational Process Performance (OPP)
Configuration Management (CM)	Organizational Training (OT)
Decision Analysis and Resolution (DAR)	Project Monitoring and Control (PMC)
Integrated Project Management (IPM)	Project Planning (PP)
Measurement and Analysis (MA)	Process and Product Quality Assurance (PPQA)
Organizational Innovation and Deployment (OID)	Quantitative Project Management (QPM)
Organizational Process Definition (OPD)	Requirements Management (REQM)
Organizational Process Focus (OPF)	Risk Management (RSKM)

modification. Each constellation will, however, add some PA components to reflect its target environment. The shared material may contain PAs that are common to more than one constellation but are not necessarily used by all constellations. We will see in Chapter 6 that the foundation concept is extremely convenient when an organization attempts to implement multiple frameworks. Variations in an organizational set of standard processes may be minimized by anchoring process definitions to the foundation.

The CMMI-DEV constellation includes two models that were in CMMI v1.1: CMMI for Development and CMMI for Development + IPPD. CMMI-DEV adds 6 PAs to the 16 core PAs:

- Product Integration (PI)
- Requirements Development (RD)
- Supplier Agreement Management (SAM)
- Technical Solution (TS)
- Validation (VAL)
- Verification (VER)

CMMI-DEV+IPPD extends CMMI-DEV with the IPPD addition. As noted earlier, the IPPD related PAs are OPD+IPPD and IPM+IPPD.

Changes from CMMI v1.1

Model changes introduced in CMMI v1.2 can be classified as those that reduced model complexity and size and those that enabled model coverage expansion. These changes are summarized in Figure 3.11. The following sections address changes

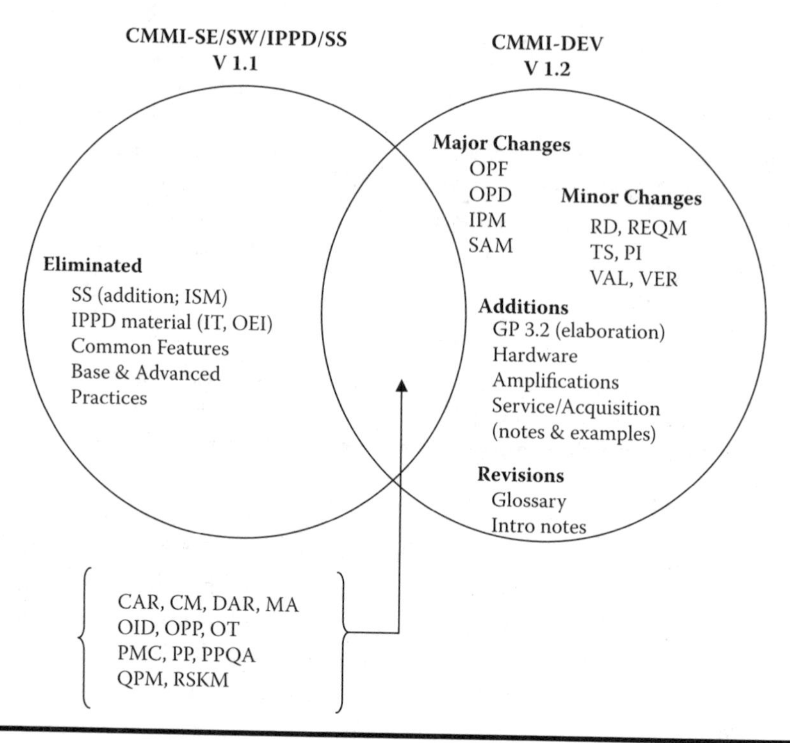

Figure 3.11 Summary of CMMI changes.

that have significant impact on model implementation. Detailed changes by PA are given in Appendix C.

The "single book" concept has now been fully adopted. CMMI v1.1 was initially published as several reports, each addressing the staged and continuous representations and various combinations of disciplines. Separate training courses were provided for each representation. The value of using both representations simultaneously (a *constagedeous* approach) has become clearer over time. Now, if desired, representation-specific content may be chosen by selecting model text marked "continuous only" or "staged only." Discipline specific amplifications may be found for hardware engineering, systems engineering, or software engineering. Text for IPPD additions are similarly marked in the body of the text.

Process Areas Eliminated

One of the major changes introduced in CMMI v1.2 was the elimination of the Supplier Sourcing discipline and the Integrated Supplier Management (ISM) PA. However, two ISM SPs—Monitor Selected Supplier Processes and Evaluate Selected Supplier Work Products—were added to SG 2 in the SAM PA.

Another major change involved the simplification of the IPPD material. Two IPPD-specific PAs, Organizational Environment for Integration (OEI) and Integrated Teaming (IT), were eliminated. Some material from those process areas was rewritten and moved to the IPPD addition's SGs in the OPD (SG 2 Enable IPPD Management) and IPM (SG 3 Apply IPPD Principle) PAs.

Process Area Improvements

In addition to the IPPD-related improvements to the OPD and IPM PAs, the OPF PA also underwent significant changes through the addition of a new specific goal, SG 3 Deploy Organizational Process Assets and Incorporate Lessons Learned. This SG adds new SPs: Deploy Standard Process and Monitor Implementation.

Through the revised OPF and IPM PAs, CMMI v1.2 emphasizes the importance of establishing, implementing, and deploying processes throughout the life of the project. The importance of applying the organization's standard processes to the early planning and requirements definition activities during startup is particularly noted. OPF SP 3.2 states "Deploy the organization's set of standard processes to projects at their startup and deploy changes to them as appropriate throughout the life of each project" and IPM SP 1.1 was changed to read: "Establish and maintain the project's defined process from project's startup through the life of the project."

REQM SP 1.4 was clarified by eliminating the reference to project plans and now reads "Maintain bidirectional traceability among the requirements and work products." The definitions of traceability and bidirectional traceability were improved.

The TS PA underwent two major changes. First, the CMMI v1.1 TS SP Evolve Operational Concepts and Scenarios was incorporated into RD SP 3.1 Establish Operational Concepts and Scenarios. Second, the CMMI v1.1 SAM SP Review COTS Products was incorporated into TS SP 1.1 Develop Alternative Solutions and Selection Criteria in CMMI v1.2 as a subpractice "Identify candidate COTS products that satisfy the requirement."

Added Work Environment Coverage

CMMI v1.2 adds two specific practices to describe processes that enable organizations and projects to benefit from using common tools, environments, and training, and highlight the importance of considering workplace issues such as ergonomics, health, and safety. The added practices are OPD SP 1.6 Establish Work Environment Standards and IPM SP 1.3 Establish the Project's Work Environment.

Added Hardware Amplifications

Amplifications were used in CMMI v1.1 to provide information relevant to the Systems Engineering, Software Engineering, IPPD, and Supplier Sourcing disciplines. In CMMI v1.2 Hardware Engineering amplifications were added to the model text where appropriate and Supplier Sourcing amplifications were eliminated.

Advanced Practices Eliminated

The continuous representation of CMMI v1.1 distinguished a limited number of "base" and "advanced" SPs for engineering PAs. This was a carryover from the systems engineering CMM and the ISO 15504 standard. This structure accommodated the achievement of CL 1 through less sophisticated approaches than expected at higher capability levels (using both base and advanced practices) or deferred the practice until higher capability levels were reached (advanced practices only). The staged representation had no such distinction. In CMMI v1.2, the concept was eliminated. Base and advanced SPs were combined into single practices for the following:

- RD SP 1.1 Elicit Needs (retained SP1.1-2 wording)
- RD SP 3.5 Validate Requirements (retained SP 3.5-1 wording)
- TS SP 1.1 Develop Alternative solutions (retained SP1.1-1 wording)
- TS SP 2.3 Design Interfaces Using Criteria (new wording)

The advanced practices retained as SPs in CMMI v1.2 are:

- REQM SP 1.2 Obtain Commitment to Requirements
- REQM SP 1.4 Maintain Bidirectional Traceability of Requirements
- RD SP 3.4 Analyze Requirements to Achieve Balance
- TS SP 1.2 Select Product Component Solutions
- TS SP 2.2 Establish a Technical Data Package
- TS SP 2.4 Perform Make, Buy, or Reuse Analyses
- PI SP 1.2 Establish the Product Integration Environment
- PI SP 1.3 Establish Product Integration Procedures and Criteria
- VER SP 1.2 Establish the Verification Environment
- VER SP 1.3 Establish Verification Procedures and Criteria
- VER SP 2.3 Analyze Peer Review Data
- VER SP 3.2 Analyze Verification Results
- VAL SP 1.2 Establish the Validation Environment
- VAL SP 1.3 Establish Validation Procedures and Criteria

With this change, the wording of GP 1.1 was changed from "Perform Base Practices" to "Perform Specific Practices."

Generic Practice Changes

The changes to GPs may be classified as minor except for the addition of elaborations to GP 3.2 Collect Improvement Information. In some cases, the GP text was modified to reflect model changes introduced in CMMI v1.2, such as elimination of base practices. GP 2.6 Manage Configurations was changed from requiring work products to be under the "appropriate levels of configuration management" to requiring the "appropriate levels of control." A few GP descriptions were improved. For example, GP 2.4 added "authority" to its description, GP 2.9's informative text notes the need to evaluate work products, and GP 5.2 points out that root causes may be beneficially analyzed even in the absence of quantitatively managed processes.

The CMMI v1.2 material relating PAs to the GPs they support was greatly expanded, showing that some SPs will support, fully or in part, corresponding GPs. An understanding of those relationships may lead to more efficient and effective institutionalization of the organizational processes.

Common Features Eliminated

In the staged representation of CMMI v1.1, GPs were grouped into categories known as common features. This was a leftover from the CMM for Software where common features were used to group the prerequisites for Key Process Area (KPA) institutionalization. Common features were never used in the continuous representation. CMMI v1.2 has eliminated common features, thus allowing an identical approach to be taken in both representations.

Glossary Changes

The CMMI Glossary underwent several major changes:

- Some new definitions were introduced, such as addition, amplification, project startup, and service
- Some definitions were revised, such as acquisition, appraisal, requirements traceability, and traceability
- Some definitions were eliminated, such as advanced practices, common features, and solicitation package

Other Changes

Other changes to the model may be considered minor but add to its clarity and usability. For example, the text defining and explaining generic goals and generic practices was moved to immediately precede the presentation of the PAs in Part Two of the model. Some editorial changes were also introduced to the model to further clarify the text and eliminate ambiguity.

Effect of CMMI v1.2 Changes on Process Improvement

Table 3.7 summarizes the impact of the CMMI v1.2 changes on process improvement, listed by capability or maturity levels.

Summary

This chapter has presented an overview of CMMI v1.2 including its architecture, components, and representations. Changes from CMMI v1.1 have incorporated lessons learned and have simplified and clarified the model. The concept of constellations allows the core concepts of the CMMI foundation to be leveraged when extending the model from development to acquisition and services.

Table 3.7 Effect of CMMI v1.2 Changes on Process Improvement

Capability/ Maturity Level	*Description of Change*	*Assessment of Impact*	*Characterization of Impact*
CL 1	Base and advanced practices	Eliminated	None
ML 2	Introduction of two new SPs from ISM	May not be applicable if SAM deals with COTS products	If COTS, minor; otherwise, major
	GP 2.4	Added "authority" wording	Minor
	GP 2.6	Changed from "levels of CM" to "levels of control"	Minor (helpful)
ML 3 (without IPPD)	Addition of work environment in OPD and IPM	New	Major
	New goal in OPF	More streamlined implementation	None to minor
	OPF SP 3.2, IPM SP 1.1	Emphasis on process implementation and deployment throughout the life of the project	Major
	Advanced practices in all Engineering PAs	Eliminated	None
	REQM SP 1.4	Reference to plans eliminated; implementation streamlined	None
	Operational concepts moved from TS to RD	Eliminated in TS	None
	GP 3.2	Elaborations added; informative material improved	None
ML 3 (with IPPD)	New and revised specific goals and practices in OPD and IPM	Material is streamlined	If already implemented using CMMI v1.1, minor; for new implementation, none
ML 4	None	None	None
ML 5	None	None	None

Chapter 4

ISO Standards

Introduction

In this chapter we will summarize the content and structure of the following ISO standards for quality management, software engineering processes, systems engineering processes, and information technology service delivery:

- ISO 9001:2000, *Quality management systems – Requirements* (ISO 2000b)
- ISO 90003:2004, *Guidelines for the application of ISO 9001:2000 to computer software* (ISO 2004a)
- ISO 15288:2008, *Systems and software engineering – System life cycle processes* (ISO 2008a)
- ISO 12207:2008, *Systems and software engineering – Software life cycle processes* (ISO 2008b)
- ISO 20000:2005, *Information technology – Service management* (ISO 2005a)

ISO, the International Organization for Standardization, was formed in the mid-1940s with the goal of unifying international industrial standards. ISO has published more than 16,000 standards, most of which are narrowly focused on specific technical criteria. In contrast to the bulk of ISO standards, ISO 9000:2000, *Quality management systems – Fundamentals and vocabulary* (ISO 2000a), probably the best known ISO standard, is a generalized standard for quality management systems. A quality management system defines the activities an organization performs to ensure that its products and services meet the needs and expectations of its customers. As such, ISO 9000 is intended to be applicable to organizations of all sizes and types, no matter what types of products and services those organizations provide.

ISO 9000, shown in Figure 4.1, represents a family of standards, which form a consistent, core set. ISO 9000 provides a set of common definitions used throughout the family of standards and establishes a number of quality management principles that guide the other standards. Taken together, these principles take a systems engineering view that is significantly different from earlier, compliance-oriented versions.

The fundamental principles established in ISO 9000 are detailed in the requirements of ISO 9001, *Quality management systems – Requirements* (ISO 2000b). ISO 9001, discussed in greater detail later, describes the general requirements for a quality management system, establishes management responsibility, and defines principles for resource management, product realization, and measurement.

ISO 9004:2000, *Quality management systems – Guidelines for performance improvements* (ISO 2000c) elaborates on each of the ISO 9001 requirements. It follows the structure of ISO 9001 and provides guidelines and interpretations for each section to support continual process improvement. It emphasizes the process approach required by ISO 9000 and reinforces the importance of following quality management principles. Although ISO 9004 elaborates the requirements of ISO 9001, it does not provide very much specific and detailed guidance for improving processes.

The three core standards—ISO 9000, 9001, and 9004—are supplemented by additional standards to assist in their application in specialty fields. For example, ISO publishes guidelines for the application of ISO 9001 in education (IWA 2:2003) and in local government (IWA 4:2005). Of interest to us here is ISO 90003:2004, *Guidelines for the application of ISO 9001:2000 to computer software* (ISO 2004a).

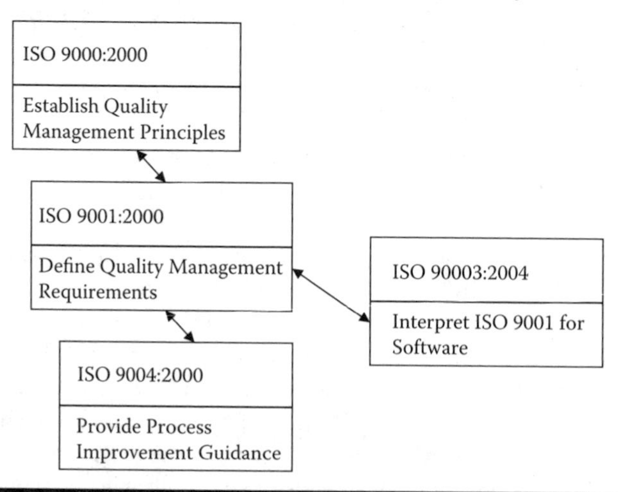

Figure 4.1 ISO 9000 family.

ISO 90003:2004 replaces ISO 9003-3:1997, which was a companion to the ISO 9001:1994 standard. Like ISO 9004, it follows the structure of ISO 9001, adding commentary and interpretation to the requirements. ISO 90003 is not itself a certification standard but rather is intended to be a useful guide. We will discuss ISO 90003 later in this chapter. ISO 90003 includes references to many other ISO standards, in particular to ISO 12207:1995 (ISO 1995), which has been recently superseded.

ISO 15288:2008, *Systems and software engineering – System life cycle processes* (ISO 2008a) provides a framework for developing and maintaining systems. This standard also addresses the complete life cycle and may be used by acquirers and suppliers, projects, and organizations. When ISO 15288 processes are applied to software engineering activities, ISO 12207 is referenced.

ISO 12207:2008, *Systems and software engineering – Software life cycle processes* (ISO 2008b) identifies the processes that are used to describe software life cycles, from initial concept through retirement. The processes are intended to address the needs of both the suppliers and providers or software systems and service.

Finally, ISO 20000:2005, *Information technology – Service management* (ISO 2005a) presents an integrated process approach to the delivery of managed information technology services. As we will see, the process approach, the definition of certain specific requirements, and the adoption of a Plan–Do–Check–Act (PDCA) methodology in this standard have much in common with the other standards with which we are concerned.

While all of these standards have some clear similarities and often include explicit cross-references, they each have their own characteristics, definitions, and points of view. For example, ISO 90003 is structured to follow ISO 9001, but frequently references ISO 12207. ISO 15288 references ISO 12207, but until the most recent revision, the two standards grouped processes differently and used different terminology. ISO 20000 is structured to address information technology service management rather than following the development, operation, and maintenance life cycle processes addressed in ISO 15288 and ISO 12207. Furthermore, revisions or amendments to the standards are usually done independently and asynchronously. This can lead to inconsistencies and confusion when an organization attempts to integrate and use them all. We will return to this issue later.

ISO 9001:2000, Quality Management Systems – Requirements

The ISO 9000 family of standards is organized around the quality management principles shown in Table 4.1. These principles are designed to work together and support each other. For example, the ISO 9000 family places a strong emphasis on understanding and satisfying customer needs. The customer defines the requirements

Table 4.1 Quality Management Principles

	Name	*Description*
1	Customer focus	An organization must understand its customers and their needs. Satisfaction of requirements may not be sufficient.
2	Leadership	Leadership provides clear vision, goals, and objectives; creates shared values; and encourages staff contributions.
3	Involvement of people	People must be involved in decision making and encouraged to take pride in their work.
4	Process approach	In any organization, many processes interact during the product life cycle, and the work products produced by one process may affect the downstream processes. The effects of process interactions on the final product, customers, and other stakeholders must be understood.
5	System approach to management	Every organization is a system of interrelated and interacting processes. The organization must be able to analyze its processes and their interactions and make corrections when required.
6	Continual improvement	Continual improvement is a permanent objective. Continual improvement can reduce variation and waste and enable quick reaction to opportunities.
7	Factual approach to decision making	Collection, analysis, and use of measurements are requirements of a process-based approach.
8	Mutually beneficial supplier relationships	An organization and its suppliers depend on each other. Effective relationships allow all parties to benefit.

for the product, including its expected quality. Customer requirements, explicit or implicit, are transformed and implemented via the product realization process, which includes verification that the product meets the customer's requirements and validation that it meets the customer's needs and expectations. Maintaining a customer focus is strongly supported by involving the organization's staff in decision making, developing an understanding of the interaction of processes, and taking a systems approach to managing those processes. Similarly, it is easy to see how involvement of people who take pride in their work, coupled with the systematic use of measurements, effectively supports continual improvement efforts.

The broad principles established in ISO 9000 are detailed in the requirements of ISO 9001, *Quality management systems – Requirements*. The introductory material in ISO 9001 reinforces the importance of the process approach, illustrates the relationships among the major sections of the standard, and recommends the use of

Table 4.2 ISO 9001:2000 Quality Management System Requirements

Section	*Title*
1	Scope
2	Normative reference
3	Terms and definitions
4	Quality management system
5	Management responsibility
6	Resource management
7	Product realization
8	Measurement, analysis, and improvement

the PDCA cycle for continual improvement. As shown in Table 4.2, sections 1, 2, and 3 of the standard define the scope, references, and terminology, and the specific requirements for a quality management system are set forth in sections 4 through 8 of the standard. The key provisions of sections 4 through 8 are discussed next.

Quality Management System

Section 4 of ISO 9001 contains the essential requirements for establishing, documenting, implementing, maintaining, and continually improving a Quality Management System (QMS). A QMS is not concerned with the requirements associated with a particular product or service. Rather, it is concerned with establishing organizational goals and objectives, defining policies and procedures needed to achieve those goals and objectives, and using those policies and procedures to operate the organization. Implementation of the requirements of this section of the standard gives management the opportunity to develop an organization built around customer satisfaction. The framework for the QMS, which is established here, provides the foundation for implementation of the rest of the standard.

An organization must document its quality objectives and policies, define and document the procedures needed for implementation of the QMS, and establish procedures for control of documents and records. The extent of mandatory documentation required by the standard is very small; only six documented procedures are explicitly required.

Decisions on the depth and breadth of QMS documentation are left to each organization and depend, for example, on the size of the organization, the scope of work performed, and the complexity of the organization's processes. Small organizations may collect all QMS documentation in a single document. Large organizations may develop multiple documents and hierarchies of quality manuals, covering, for example, local, national, and international requirements. No matter how the documents are organized, they need to be controlled and kept current, and

Table 4.3 Typical Quality Manual Table of Contents

Section	Title	Contents
1	Scope	Define the organization to which the manual applies
2	Review, Approval, and Revision	Evidence of review, approval, revision status, and date of the Quality Manual
3	Quality Policy and Quality Objectives	Quality policy and quality objectives can be included in the Quality Manual or in a separate document. Quality goals are measurable entities, which satisfy the objectives.
4	Organization, Responsibility, and Authority	The structure of the organization should be defined. Responsibility, authority, and interactions may be indicated, for example, by means of organization charts or job descriptions.
5	References	List of documents referenced in the text.
6	Quality Management System Description	The description of the QMS and its implementation in the organization is provided. Processes and their interactions are described or referenced. This section also contains descriptions of, or references to, documented procedures. To facilitate reviews and audits, most Quality Manuals list processes in the same order as the ISO 9001 standard sections.
7	Appendices	Supporting material may be included here.

the records that result from process execution must be maintained to show that the QMS has been followed.

The documentation describing an organization's QMS is captured in a Quality Manual. ISO/TR 10013:2001, *Guidelines for quality management system documentation* (ISO 2001a) promotes the development of documentation that supports the process approach of the ISO 9000 series. A typical Quality Manual table of contents, suggested but not mandated by ISO 10013, is shown in Table 4.3.

Management Responsibility

Virtually every study, report, and anecdote finds that a key ingredient in the successful introduction of new processes or improvement of existing processes is the commitment and support of management. Section 5 of ISO 9001 addresses these topics.

Commitment and support mean more than just assigning responsibility to a member of the management team and providing some funding. True leadership comes from establishing organizational values, leading by example, recognizing employee's work, and being actively involved in implementation of the organization's processes.

Management commitment is initially established by defining and disseminating the organization's quality objectives and quality policy. To do this, adequate resources must be made available including funding, properly trained people, and appropriate tools, such as support software and measurement equipment.

When establishing the organization's quality objectives, the needs of the organization, anticipated future needs, market direction, feedback from measurement of the performance of current products and processes, and benchmarks against the industry and competitors must be considered.

Implementation of an effective QMS requires an understanding of the organization's stakeholders. These stakeholders, or interested parties, include:

- Customers
- End users
- Employees
- Suppliers
- Stockholders

It is the responsibility of management to ensure that the stakeholders are identified and that their needs and expectations are understood. A more difficult—but necessary—task for management is balancing the needs and desires of stakeholders. Relationships among the stakeholders must be established, prioritized, and maintained. Although customer and stakeholder satisfaction is an important theme throughout the ISO 9001 standard, it does a commercial organization no good to satisfy those stakeholders and then have to go out of business!

Quality management system planning thus requires consideration of:

- Satisfaction of customer needs and desires
- Quality objectives
- Definition and communication of responsibility and authority
- Regulatory requirements
- Measurement of process and product performance
- Interaction among QMS processes

Part of establishing responsibility and authority for various activities throughout an organization is ensuring that staff members are kept aware of the importance of their roles and are encouraged to take pride in their work and contribute to the organization's success.

Finally, management needs to regularly review the performance of the QMS. These reviews are conducted not merely to verify compliance, but to ensure that the QMS processes remain effective, efficient, and appropriate for the organization, and to identify improvement opportunities where there are shortfalls. Information reviewed includes customer feedback, performance measurements, supplier performance, improvement recommendations, and strategic plans.

Resource Management

The world's finest QMS is of little use if the resources needed for its implementation are not provided. Resources, including people and the environment in which they work, are discussed in section 6 of ISO 9001. This section provides strong support for the ISO 9000 principle that the involvement of people is an important part of a successful quality management system.

An organization and its people have a symbiotic relationship. The organization must identify its long- and short-term objectives and then identify the skills and environments needed to meet those objectives. People in the organization contribute to both the definition of objectives and identification of effective means to satisfy those objectives.

Management must stay aware of the needs of people, recognize achievement, encourage innovation, and maintain communication. One activity that supports these goals is providing effective training to ensure that people are competent to perform their jobs. Competence means that not only are people aware of the work required, but they can demonstrate the ability to perform that work. Thus, training must be provided and the effectiveness of that training must be evaluated.

Among the other resources that must be provided and managed are the physical infrastructure and the work environment. The infrastructure and environment include:

- Buildings
- Equipment
- Communications facilities
- Data management
- Heating, cooling, and lighting
- Safety and operational rules
- Tools
- Support services

Product Realization

Section 7, product realization, is the largest section of ISO 9001. This section requires that the QMS define the processes used to develop and deliver products and services. The principles of process engineering and systems engineering are strongly supported.

The processes defined in the QMS are selected, tailored, and documented in a Quality Plan for use when realizing a particular product or service. Planning for product realization includes:

- Determination of product requirements and quality goals
- Identification of processes
- Understanding process inputs and outputs, and their relationships
- Coordination of processes
- Training
- Verification, validation, measurement, and analysis

The selection of product realization processes must consider the organization's quality objectives, the product-specific quality objectives, the product's intended end use, and verification and validation requirements.

When developing products, organizations must determine and maintain customer-specified requirements. In addition, implied requirements, legal and regulatory requirements, and self-imposed requirements must be incorporated into the product realization process. Records of reviews, assumptions, changes, and decisions must be maintained and communicated to the customer.

Design and development activities include the selection of life cycle models, determination of the steps needed for each life cycle phase, and designation of responsibility for each step. The organization must define both the required process inputs and the planned outputs. Inputs are provided from sources external and internal to the product development organization. Outputs should be defined to allow verification that the product meets requirements and that the process outputs are suitable as inputs to subsequent processes. Changes to requirements and design will almost certainly be made during execution of design and development processes, and those changes must be controlled, recorded, and communicated.

Systematic reviews of design and development should be planned and conducted. These reviews address:

- Progress against plans
- Verification and validation
- Problems and risks
- Product and process performance measures
- Improvement opportunities

Management is responsible for ensuring that appropriate purchasing processes are developed and implemented. These processes include systematic selection of suppliers using defined criteria, control of the supplier, and verification that the purchased products meet requirements. The extent of supplier control needs to be adjusted based on the significance of the purchased product, on subsequent design and development activities, and on the final product.

Several clauses in this section of the standard deal with establishing the infrastructure and supporting processes needed for effective product realization. These include ensuring that product information, procedures, work instructions, equipment, and measurement devices are made available. Mechanisms must be in place so that raw material, parts, work in progress, finished goods, customer equipment, and data can be identified and safeguarded.

Finally, devices used for monitoring products and processes must be controlled. Control includes calibration, maintenance of calibration status, and prevention of inappropriate adjustments to the equipment or results.

Measurement, Analysis, and Improvement

Section 8 of ISO 9001 discusses the use of measurements, an important part of every well-established engineering discipline, to monitor and improve products and the QMS. Measurement and analysis allows organizations to evaluate products, determine process capability and actual process performance, and take any needed actions. Measurements allow an organization to demonstrate that product requirements have been met and provide the insights needed for improvement of the QMS.

In support of the ISO 9000 principle of customer focus, ISO 9001 requires the development and implementation of processes for monitoring customer satisfaction. Significantly, this includes collecting information on the customer's *perception* that requirements have been met and does not solely rely on actually meeting contractual requirements. Thus, these measurements may include acceptance test data, customer surveys, changes in order volume, and explicit positive and negative feedback.

Another important requirement of this section discusses the need for an organization to conduct planned internal audits of its QMS. These audits are used to determine conformance of the activities actually performed to the QMS, conformance of the QMS to other standards (such as ISO 9001), and to identify opportunities for improvement. Identified nonconformances should be corrected. Related requirements in this section discuss monitoring and measuring the QMS processes.

Products developed by implementing the product realization processes must also be monitored and measured to verify that both the final product and intermediate work products satisfy their requirements. Records of these measurements must be maintained.

Products that fail to satisfy their requirements must be controlled to prevent their unintended use. Procedures for detecting, controlling, and disposing of nonconforming products must be part of the QMS. Disposition of such nonconforming products may include, for example, rework, scrap, or waiver from the customer.

A necessary component of a measurement program is the analysis of the collected data. For example, data analyses are used to show status and trends in:

- Customer satisfaction
- Employee satisfaction
- Quality
- QMS performance
- Supplier performance
- Economic performance

Beyond the correction of nonconformances detected through reviews, audits, and product verification activities, management has the responsibility to actively and continually seek to improve the QMS. Therefore, action must be taken to prevent the recurrence of nonconformances by eliminating their causes and also eliminating the causes of potential future nonconformances.

ISO 90003:2004, Guidelines for the Application of ISO 9001:2000 to Computer Software

As previously noted, ISO 9001 is a broadly applicable, generalized quality management standard. Software-specific interpretations of ISO 9001 are provided by ISO 90003:2004. ISO 90003 is not itself a certification standard; it is intended to be a useful guide whether or not the organization seeks ISO 9001 registration. It is independent of technology, life cycle models, development processes, and organizational structure.

ISO 90003 includes numerous references to other ISO standards, explicitly showing cross-references to 19 other ISO standards. It is especially reliant on ISO 12207:1995 (ISO 1995) and ISO 12207/Amendment 1:2002 (ISO 2002b) for definitions and elaborations that support and complement implementation. The newly published ISO 12207:2008 and ISO 15288:2008 will have a great impact on the usability of this standard until it is revised to reflect their existence.

A significant feature of ISO 90003, in contrast to ISO 9001, is the explicit recognition of the way in which different sections of the standard are aimed at different parts of the organization. As shown in Figure 4.2, sections 4 (Quality management system), 5 (Management responsibility), and 6 (Resource management), and part of section 8 (Measurement, analysis and improvement) are primarily applicable at the organizational level but have some application at the project level. Section 7 (Product realization) and the remainder of section 8 primarily describe project level actions. This structure allows the experiences of multiple projects to be leveraged to improve the organization's processes. At the same time, it allows projects to tailor processes to meet their specific needs while taking advantage of organizational lessons learned.

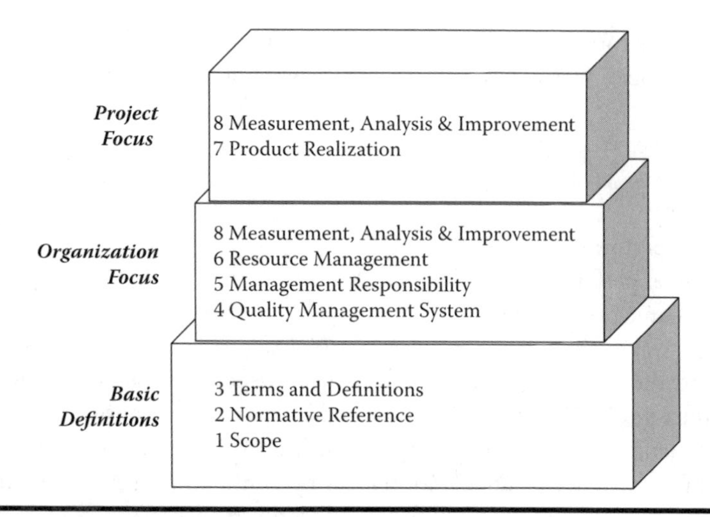

Figure 4.2 ISO 90003 focus.

Quality Management System

ISO 90003 expands on the general requirements of ISO 9001 by explicitly noting the need to identify software-related processes, including development, maintenance, and operations processes; showing where those processes appear in various life cycle models; and defining the interactions among processes. Where ISO 9001 gives only general direction for including documented procedures in the QMS, ISO 90003 suggests several items needed for software. These include:

- Processes
- Procedures
- Templates
- Life cycle model descriptions
- Descriptions of tools and techniques

Records should be created and maintained to show that requirements have been met, the QMS has been followed, and the QMS functions effectively. Decisions must be made on the format and media (e.g., paper or electronic) to be used and the duration of record retention. Typical records used to show that requirements have been met include:

- Audit reports
- Inspection reports
- Test results
- Problem reports
- Change requests and requirements changes

Records showing use of the QMS include:

- Size, cost, schedule, and resource estimates
- Changes to estimates
- Minutes of meetings and reviews
- Rationale for selection of life cycle model and development methodologies

Management Responsibility

Little additional detail is provided for interpreting this section of ISO 9001 for software. Management is responsible for ensuring that the QMS addresses software-specific issues. Examples indicated by ISO 90003 include:

- Identification of life cycle models appropriate to the work done in the organization
- Guidance on the tailoring of development methods
- Definition of expected development work products, such as plans, specifications, software, and operations manuals
- Specification of tools and engineering environments

Management also needs to consider process performance and product conformity. This can be done by using the results of software process assessments and software product evaluations. These activities are discussed in section 8 of the standard.

Resource Management

This section expands on the ISO 9001 discussion of training needs with respect to software development, maintenance, and operations. Training should address the technologies used for software in the organization, such as specific languages, tools, and development methods. The organization's technology choices need to be monitored and revised as appropriate. Skill requirements and training plans consequently need to be kept current.

Training need not be limited to classroom instruction. Other methods such as mentoring, on-the-job training, independent study, and computer-based instruction may be equally useful, easier to schedule, and more cost-effective.

This section of the standard also makes note of specific examples of infrastructure support tools needed in software organizations. These support tools, which should be kept under the appropriate levels of control, may include:

- Tools for analysis, design, coding, and testing
- Support tools for configuration management, measurement and analysis, and quality assurance

- Project management and presentation tools
- Software libraries
- Network support tools

Product Realization

The bulk of software-specific interpretations and guidance provided by ISO 90003 appears in this section.

Planning of Product Realization

Each software development or maintenance project needs to select the life cycle that best fits its needs and constraints. For example, safety- or security-critical applications will not be developed in the same way as software for Web-based advertising. A waterfall life cycle model may be the best fit for well-defined, precedented systems, whereas an evolutionary or agile development method may be best for systems with vaguely defined requirements.

A significant part of quality planning for software systems involves tailoring the QMS to fit the particular needs of a specific project. An expectation of the ISO 90003 standard is that those sections of a quality plan that apply to a particular development stage should be complete prior to starting that stage. In addition, the quality plans should be reviewed and revised during the course of product realization. ISO 90003 treats quality planning and development planning separately, while noting that the related ISO 12207:1995 standard addresses these activities together. In any event, the quality plan may be separately documented or packaged with other documents.

Topics addressed in a project-level quality plan include:

- Product quality requirements
- Process quality requirements
- Reference to development plans
- Identification of processes, procedures, and work instructions used by the project
- Description of the tailoring of the QMS
- Identification of the project life cycle
- Identification of project tools, languages, conventions, and standards
- Entry and exit criteria for each phase of the project
- Plans for reviews, and verification and validation activities
- Measurement and analysis plans
- Configuration management and data management plans
- Training plans

Customer-Related Processes

The customer-related processes section addresses determination of product requirements, review of requirements, and communication with the customer. Software may be developed for a variety of customers and uses, such as:

- For a single customer under a contractual agreement
- Broadly available for a market segment
- For internal organizational use
- As a component of a larger system

Requirements definition and development are needed for each of these customer types and environments, although the methods used will generally differ. For example, under a contractual agreement, a specification may be provided but the requirements in that specification may not be fully defined or articulated. In cases where the product is intended for use in a broad market segment, no particular customer exists, so customer surrogates, such as surveys or competitive analyses, must be substituted.

Furthermore, product requirements must also include any additional requirements defined by the organization. Here, ISO 90003 doesn't provide any specific guidance but rather refers to a number of additional ISO standards, including ISO 12207:1995 and ISO 12207:1995/Amendment 1.

The organization must review the product requirements before committing to implement them. Among other topics, reviews will typically address:

- Requirements feasibility
- Requirements for product reliability, maintainability, portability
- Hardware and software platforms
- Interfaces
- Life cycle and process selections
- Customer imposed life cycles and processes
- Risks
- Resource requirements and availability

Customer communications are clearly needed as part of the requirements definition and at the end of product development, but continuing communication throughout the development cycle is also important. Communication during product development allows for customer feedback, management of risks, and an understanding of changes to the customers needs and requirements. Mechanisms for customer communications should be defined as part of the quality plan.

Design and Development

This section addresses planning for design and development, specification and review of design and development inputs and outputs, verification and validation of outputs, and control of changes.

The discussion of design and development planning emphasizes the need for disciplined execution of the plan so that verification and validation activities are not the only methods used to detect the need for corrective actions. Quite a few topics that need to be addressed during planning are identified by the standard. Among these are:

- Identification of the life cycle model, methods, tools, and techniques
- Identification of development facilities
- Definition of configuration management practices
- Planning the activities of analysis, design, development, implementation, integration, testing, and support
- Project resource planning
- Definition of interfaces with other groups
- Identification of related planning activities
- Risk analysis
- Development of a work breakdown structure
- Schedule definition
- Plans for document control

Design and development activities are driven by requirements. Requirements may be defined explicitly in specifications, but they are also developed through subsequent analysis and consideration of performance, quality, interfaces, system design, and statutory constraints. In any event, requirements should be reviewed for consistency, ambiguity, verifiability, realism, and omissions. Requirements reviews often include customer representatives. The ISO 90003 standard refers the reader to ISO 9126-1:2001 (ISO 2001b) for a more extensive discussion of software quality requirements and characteristics.

The outputs of the design and development process must satisfy the input requirements. The output of design and development is generally the developed product but may also include design documentation, source code, test documentation, training material, and user documentation. A reference to ISO 12207:1995 is provided for further information on the results of performing design and development activities.

Design and development activities and results should be reviewed at appropriate times. Determination of the meaning of *appropriate* depends, for example, on quality requirements, product complexity, risks, and contractual constraints. It is expected that reviews are performed in accordance with the plans previously established for design and development, addressing:

- The subject of the review (such as design documentation or prototype demonstration)
- Groups to be included in the review (such as marketing or the customer's representative)
- Actions to be taken in preparation for the review
- Expectations for conduct of the review
- Definition of success criteria
- Specification of the data to be produced (such as meeting minutes or action item lists)

Verification of work products takes place throughout design and development to ensure that the input requirements have been satisfied, whereas validation is performed to ensure that the software meets operational needs and requirements. Verification and validation share some of the same techniques, such as test and demonstration, but they serve different purposes. Validation in the intended operational environment is generally desirable but is not always possible. For example, validation under operational conditions may not be feasible for software designed to operate on the moon or during earthquakes.

Finally, changes during design and development must be controlled, generally through configuration management activities. When changes are made, related work products, such as design, code, and documentation, must be kept consistent with one another.

Purchasing

ISO 90003 lists a broad array of items under the purchased products banner. These include:

- COTS software
- Custom software
- Free software
- Outsourced product development and support services
- Contracted staff
- Hardware
- Documentation
- Training

For each of these areas, the purchased products must be controlled and verified to ensure that the product requirements have been met. In practical terms, complete verification may not always be possible. This would be the case, for example, when the use of third-party software has been mandated by the customer or when commercially available word processing software is purchased.

Production and Service Provision

For software, the production and service provision may be interpreted as:

- Configuration management activities, including identification, traceability, and build and release activities
- Delivery and installation activities
- Operations, maintenance, and support activities

Configuration items must be identified by name and version and by specifying when they are brought under configuration control. The appropriate versions of each configuration item must be identified to specify the particular version of each complete product. The traceability of components to the complete product should be maintained throughout product realization.

Software product delivery may be via physical media or through electronic transmission, but in either case, the software must be protected from damage. This can include protection of the physical media and measures to prevent alteration of the contents, for example, through virus checking or data encryption. Installation steps will vary depending on, for example, product complexity and end-user sophistication. Installation may require expert configuration and tuning by the development organization or may be done by the customer using an installation script.

ISO 90003 has little to say about operations, other than to note the need to plan for help desk activities, disaster recovery, and security. Maintenance processes should be established and should specify the scope of the maintenance and support activities, configuration management plans, and the maintenance of records.

Throughout design, development, production, and service activities, the integrity of customer property must be maintained. Customer property may include data, hardware, software, tools, development and test environments, and proprietary information. Plans for controlling and protecting customer property are often covered by the organization's configuration management plans.

Control of Monitoring and Measuring Devices

Interpretation of this section of ISO 9001 for software is something of a forced fit. The interpretation presented by ISO 90003 is to consider tools, facilities, and techniques to be the monitoring and measuring devices for software. Thus, tools are controlled by placing them under configuration management. Tools and techniques are "calibrated" and readjusted by considering their effect on product quality and improving them as needed.

Measurement, Analysis, and Improvement

Measurement and analysis is performed to effectively manage processes and to demonstrate product quality.

Monitoring and Measurement

For software, quality measures related to customer satisfaction emerge from analysis of help desk calls, postrelease fixes needed, and customer feedback on quality-in-use metrics such as those defined in ISO 9126-4 (ISO 2001c).

Organizations perform internal audits to help ensure that their QMS is effective. Software organizations may designate selected projects for audit and evaluate compliance of project plans to the organization's QMS and compliance of the project to its plans. The audits are conducted at various points during the product's life cycle.

Selected processes may be chosen to be monitored and measured. Measures may include, for example, planned and actual process cost, planned and actual process duration and schedule, and the quality characteristics of the process output. Similarly, the quality of selected products will be measured. Typical product measures include functionality, portability, and efficiency.

Control of Nonconforming Product

Nonconforming products are controlled by eliminating the nonconformity, ensuring that the product is not used, or by reaching agreement with the customer to allow its use. For software, these steps are largely implemented through the use of configuration management and defect management processes. Software defects must be investigated and corrected. The resulting corrected work products are then entered into the configuration management system and the appropriate baselines are updated.

Analysis of Data

Analysis of data for software will generally examine test reports, peer review results, review comments, action items, and customer feedback.

Improvement

Without providing very much elaboration, ISO 90003 suggests establishing an improvement process to support the ISO 9001 requirement for continual improvement. Corrective action is largely handled through configuration management and it is noted that process assessments may be helpful in data collection leading to preventive action. References for further information are provided pointing to ISO 12207:1995, ISO 12207/Amendment 1, and ISO 15504.

Harmonization of Standards

As we have noted, standards are published at different times and may include cross-references to the then-current versions of other standards. As standards and their guidebooks are amended and revised, those cross-references may become obsolete. Furthermore, even closely related standards often use different terminology, have different process architectures, and differ in presentation style.

For example, ISO 90003 is structured to follow ISO 9001 exactly, so there is no difficulty in aligning the two standards. On the other hand, ISO 90003 makes a considerable number of references to ISO 12207:1995 and its amendments. ISO 12207 has, in fact, undergone revision, so there will be discontinuities between ISO 90003 and the latest version of ISO 12207 until ISO 90003 is revised.

The ISO 12207:1995 and ISO 15288:2002 standards addressed closely related topics, namely, software and systems engineering life cycle processes. However, the standards used different terminology, defined different processes, addressed those processes through different architectures, and tended to use different levels of prescription. Because each standard was technically sound individually and because the user community wanted to be able to use them together, an effort was undertaken to harmonize them. The harmonization effort is being performed in two phases.

The first phase of harmonization involved bringing both standards into alignment while maintaining compatibility with their previous versions. Because the previous versions of the standards categorized life cycle processes differently, the harmonization effort created a uniform architecture where processes are categorized identically and the decomposition of processes into activities, tasks, and informative material follows the same structure. Some life cycle processes were added or renamed and some tasks were redistributed across processes to enhance interoperability of the standards. To the extent possible, a common vocabulary and a similar balance of prescriptive versus descriptive text are now used. This phase of harmonization was completed at the end of 2007.

As of this writing, the second phase of harmonization is being planned. Among the topics being considered for the integrated view of systems and software life cycle phases are:

- Greater commonality in the level of prescriptive text
- Common configuration management concepts
- Common verification and validation concepts
- Applicability to services in addition to products
- Handling of life cycle models
- Alignment with other standards

The revised ISO 12207 and ISO 15288 standards will, in the future, share a common guide to life cycle management, ISO 24748 (ISO 2008c).

ISO 15288:2008, Systems and Software Engineering – System Life Cycle Processes

ISO 15288:2008 addresses life cycle processes for systems. The standard is intended to be broadly applicable to all man-made systems and can be used over the entire life cycle, from system conception to system retirement. The standard may be applied by both acquisition organizations and supplier organizations, and may be used within a single organization and across multiple organizations. Significantly, the architecture established by ISO 15288 is held in common with ISO 12207 so that the two standards may easily be used concurrently. As we will discuss in chapter 6, this commonality can be successfully exploited when developing an overall process approach and architecture.

The processes defined in ISO 15288 are shown in Figure 4.3. Each process addressed by the standard includes the definition of the purpose and desired outcomes and identifies, at a high level, the activities and tasks addressed by each

Project Processes

- Project Planning Process
- Project Assessment and Control Process
- Decision Management Process
- Risk Management Process
- Configuration Management Process
- Information Management Process
- Measurement Process

Technical Processes

- Stakeholder Requirements Definition Process
- Requirements Analysis Process
- Architectural Design Process
- Implementation Process
- Integration Process
- Verification Process
- Transition Process
- Validation Process
- Operation Process
- Maintenance Process
- Disposal Process

Organizational Project-Enabling Processes

- Life Cycle Model Management Process
- Infrastructure Management Process
- Project Portfolio Management Process
- Human Resource Management Process
- Quality Management Process

Agreement Processes

- Acquisition Process
- Supply Process

Figure 4.3 ISO 15288 life cycle processes.

process. It does not, however, indicate any particular standards or procedures to be followed for implementation of each process. The standard may be used as a process reference model for assessment under ISO 15504, *Information technology – Process assessment* [ISO 2004b].

As we have seen with other standards, ISO 15288 notes that the identification of life cycle processes does not imply any particular life cycle model or impose an ordering on the processes. It does, however, require that appropriate life cycle models be established and that each stage in the life cycle have a defined purpose and outcome. The outcomes defined for each of the processes identified in the standard can provide guidance when establishing the ordering of processes for a specific life cycle definition. ISO 15288 presents the system life cycle processes in four groups, as shown in Table 4.4.

Agreement Processes

The Agreement processes are used to establish business arrangements between acquiring and supplying organizations. The Agreement processes are:

- Acquisition process
- Supply process

Acquisition Process

The Acquisition process is used to obtain products and services. In this process, the plan for conducting the acquisition is developed, suppliers are selected, and agreements are negotiated. Implementation of the agreement is monitored by both

Table 4.4 ISO 15288:2008 System Life Cycle Processes

Group	*Process*	*Scope*
Agreement	Acquisition	Obtain products or services
	Supply	Provide products or services
Organizational Project-Enabling	Life Cycle Model Management	Maintain the organization's policies, life cycle models, processes, and procedures
	Infrastructure Management	Manage organizational infrastructure and resources to support projects
	Project Portfolio Management	Initiate projects needed to support organizational goals and policies
	Human Resource Management	Provide qualified, competent staff to projects

Group	*Process*	*Scope*
	Quality Management	Assure products and services meet quality goals and satisfy customers
Project	Project Planning	Develop and disseminate project plans and schedules
	Project Assessment and Control	Determine project status and ensure project execution according to plans, schedules, and budgets
	Decision Management	Analyze and select decisions from among alternative choices
	Risk Management	Identify, assess, and manage risks
	Configuration Management	Maintain integrity of selected project and process outputs
	Information Management	Collect, maintain, and retrieve selected information
	Measurement	Collect, analyze, and report process and product measures
Technical	Stakeholder Requirements Definition	Identify system requirements for services needed by users and other stakeholders
	Requirements Analysis	Transform stakeholder needs to technical system requirements
	Architectural Design	Define the components and structure of a solution addressing the system requirements
	Implementation	Produce specified components that address architectural design requirements
	Integration	Assemble system components in accordance with the architectural design
	Verification	Confirm that design requirements have been met
	Transition	Prepare to provide specified services in the operational environment
	Validation	Show that the system meets the stakeholder requirements
	Operation	Use the system to provide specified services
	Maintenance	Sustain capability through corrective, preventive, adaptive, and preventive actions
	Disposal	Retire and remove the system

parties and payment or other compensation is provided for services rendered and products delivered.

Supply Process

The Supply process is the counterpart of the acquisition process. It includes a marketing activity to identify potential customers for the supplier's products and services. In the supply process, an offer is made in response to the acquirer's request. If the supplier is selected, an agreement is negotiated and implemented, products and services are delivered, and payment is received.

Organizational Project-Enabling Processes

The Organizational Project-Enabling processes, known as the enterprise processes in the previous version of the standard, provide the organizational infrastructure needed to support and control projects. With an eye toward the organization's overall business objectives, these processes are also used to determine which efforts will receive funding and other resources from the organization. The Organizational Project-Enabling processes are:

- Life Cycle Model Management
- Infrastructure Management
- Project Portfolio Management
- Human Resource Management
- Quality Management

Life Cycle Model Management

The Life Cycle Model Management process establishes organizational policies and procedures for system life cycle processes and defines the organization's standard life cycle models. The policies, procedures, responsibilities, and authorities that are established here are communicated to projects. This section also includes activities for assessing and improving the organization's processes, and makes specific reference to ISO 15504 for details on assessment activities.

Infrastructure Management

The Infrastructure Management process activities establish and maintain the resources needed to address project and organizational objectives. Resources include standards, software, hardware, tools, and facilities.

Project Portfolio Management

The Project Portfolio Management process controls the commitment of organizational funding and resources to establish and sustain selected projects. The expectations, authorities, and accountability for each project are established and resources are allocated to approved projects. The progress of each project is monitored against plans, and decisions are made to continue, change, or cancel funding.

Human Resource Management

The Human Resource Management process is used to provide projects with the skilled personnel needed to meet project objectives and maintain the competencies of the organization's staff. Human Resource Management also maintains the strategy and services needed to manage knowledge within the organization and is used to manage resource demands among projects.

Quality Management

The purpose of the Quality Management process is to assure that the organization's quality goals are achieved and that customers are satisfied. Under this process, quality policies and procedures are established and the organization's quality goals are defined. Here, ISO 15288 makes specific reference to the quality management system requirements defined in ISO 9001 and ISO 9004. Product and service quality, quality improvements, and customer satisfaction are assessed and action is taken when needed.

Project Processes

Project processes are used to define project plans, track progress against plans, and take action to direct project progress. The standard notes that these should not be considered to be the complete set of nonengineering processes needed to manage a project. These processes are:

- Project Management
  - Project Planning
  - Project Assessment and Control
- Project Support
  - Decision Management
  - Risk Management
  - Configuration Management
  - Information Management
  - Measurement

Project Planning

The Project Planning process produces the plans that define project activities, assigns roles and responsibilities, identifies resource requirements, and defines performance measures. To achieve these results, the project scope must be defined and a work breakdown structure must be developed to identify project activities. Project costs, including labor and purchased products and services, need to be estimated and a budget and schedule must be established. The approach to managing the project, measuring performance, and reviewing progress must be included in the project plans.

Project Assessment and Control

The Project Assessment and Control process is used to evaluate project status and manage project execution to keep project performance aligned with project plans. When deviations from planned performance are detected, responsible parties are notified so that corrective and preventive action may be taken. Activities performed in this process are used to evaluate the adequacy of project team competencies, the supporting infrastructure, and the status of cost, schedule, and quality performance compared to plans. The project direction may be replanned if prior estimation and planning assumptions change or are incorrect. Technical and management reviews are conducted to establish the readiness for the subsequent life cycle stages. Finally, project closeout activities, such as storage of project records, are addressed at completion.

Decision Management

During the course of any project, many decisions are needed. Certain problems—but not all—require a formal process for identifying a decision strategy, identifying alternatives, evaluating those alternatives, and selecting a course of action. For those decisions that use a formal process, records of the issue, decision, and rationale are captured to enable organizational learning.

Risk Management

The Risk Management process uses a systematic approach to identify, assess, and act upon potential events, which may negatively affect cost, schedule, quality, or technical performance. First, an overall risk management strategy that includes assignment of responsibilities and definition of risk thresholds is established. Then, risks are identified and assessed. The probability of risk occurrence and potential consequences are estimated, the identified risks are prioritized, the approaches for their management are determined, and the risks are monitored. The standard also includes an activity for evaluating and improving the risk management process itself.

Configuration Management

The integrity of project work products is established and maintained through the Configuration Management process. The process requires the definition of a strategy for controlling access, release, and changes to items in the configuration management system. Items that are under configuration control are identified and information on current configurations is maintained. Changes to configuration items must be identified, evaluated, and approved prior to implementation.

Information Management

During the course of any systems engineering effort, information in a variety of forms must be collected, maintained, and disseminated to various parties. These activities are addressed by the Information Management process. As part of this process, the organization must identify the information to be managed, define the formats and media, and establish responsibility for information collection, storage, release, and disposition.

Measurement

The Measurement process, a new addition to the latest version of ISO 15288, is focused on addressing an organization's information needs. Measures, collection procedures, and analyses addressing information needs are defined and implemented, and the results are reported to measurement users. The standard notes that this process supports the ISO 9001 measurement requirements and that detailed activities and tasks, consistent with this process, may be found in ISO 15939 (ISO 2007).

Technical Processes

The Technical processes are used to define system requirements, produce and operate the system, and ultimately to retire and dispose of the system. The Technical processes defined by ISO 15288 are:

- Stakeholder Requirements Definition
- Requirements Analysis
- Architectural Design
- Implementation
- Integration
- Verification
- Transition
- Validation

- Operation
- Maintenance
- Disposal

Stakeholder Requirements Definition

The Stakeholder Requirements Definition process begins with the identification of stakeholders, which includes users, acquirers, suppliers, developers, operators, and others. The needs, expectations, and requirements of these stakeholders are elicited and the intended use of the system is defined in terms of user interactions and operational scenarios. Constraints are identified and requirements issues, such as inconsistency and conflict, are resolved. The stakeholder requirements are reviewed to verify a shared understanding and traceability to stakeholder needs and desires.

Requirements Analysis

Whereas the Stakeholder Requirements Definition process is used to determine the user's view of the system needs, the Requirements Analysis process is used to develop the corresponding technical requirements. The Requirements Analysis process defines the functions to be performed by the system, specifies quality and performance requirements, identifies interface requirements, and determines constraints on implementation. Traceability between user requirements and system requirements is developed and maintained.

Architectural Design

The Architectural Design process uses the technical requirements developed by the Requirements Analysis process to develop a solution that will serve as the basis for system implementation. A system architecture is developed and requirements are allocated to each element of the architecture. The functions performed by each element are allocated to hardware, software, or human implementation and the interfaces between elements are defined. Traceability between the system requirements and the design elements is maintained.

Implementation

The system design developed by the Architectural Design process is transformed into realized system elements by the Implementation process. A strategy is developed to define the technologies to be used, and, following that strategy, hardware is fabricated, software is coded, operators are trained, and system elements are inspected and tested.

Integration

The system elements developed by the Implementation process are assembled into a complete system by the Integration process. Using an integration strategy that defines the sequence in which system elements are to be assembled, components are brought together. The Verification and Validation processes are used repeatedly to ensure that the system performs as planned. The status of integration activities, including discrepancies, is recorded and analyzed.

Verification

The Verification process is used to ensure that the system satisfies its requirements. Here, requirements includes not only those developed by the Requirements Analysis process, but those developed by other life cycle processes such as Architectural Design and Implementation. To accomplish this goal, the Verification process is applied throughout the system's life cycle in accordance with a defined verification strategy and plan. The status of verification activities is recorded and analyzed, discrepancies are noted, and corrective actions are implemented.

Transition

Under the Transition process, a verified system is installed in its operational environment. Transition activities include development of a transition plan, site preparation, system delivery, system installation, and system activation. Installation data, including discrepancies and corrective actions, are recorded.

Validation

Whereas the Verification process is used to show that system requirements have been satisfied, the Validation process is used to show that the system meets the stakeholders' requirements. Validation activities are planned and may be conducted throughout the system life cycle. Validation early in the life cycle may be performed using paper analyses or prototypes, whereas validation at the end of development typically uses tests and demonstrations in an operational environment.

Operation

The Operation process delivers the services of the system. The operation plan defines the strategy for delivering and withdrawing services, integrating and coordinating with existing systems, scheduling services, and staffing with trained operators. System operation is monitored to make sure performance is acceptable, system failures are identified and addressed, and user satisfaction with system services is monitored.

Maintenance

The Maintenance process is used to ensure that the system continues to provide useful service. A maintenance strategy is planned and includes preventive maintenance to help ensure continued successful operation, corrective maintenance to address defects, adaptive maintenance to account for environmental changes, and perfective maintenance to introduce improvements.

Disposal

Eventually, systems need to be retired and disposed. The Disposal process addresses the definition of a disposal strategy; system deactivation; removal of system components in accordance with health, safety, and environmental constraints; and archiving of system information for potential future review.

Annexes and Guides

The ISO 15288 standard includes several annexes to assist in its interpretation and implementation. These annexes are:

- Tailoring Process (normative)—Summarizes the activities for tailoring the standard to meet the needs of an organization or project.
- Process Reference Model for Assessment (informative)—Process assessments under ISO 15504 require the use of a Process Reference Model (PRM). ISO 15288, including the purpose statement, the outcomes, and the activities and tasks for each life cycle process, satisfies the requirements of a PRM.
- Process Integration and Process Constructs (informative)—As a result of the harmonization efforts, the process constructs used by ISO 15288 and ISO 12207 are now aligned. This annex discusses the grouping of processes and the consistent structure of process definitions such as process name, purpose, and outcomes supported by activities. Activities, in turn, are supported by tasks and notes.
- Process Views (informative)—Presents the concept of organizing activities pertaining to a particular engineering discipline in one place, even though those activities may be drawn from multiple life cycle processes.
- ISO 15288 and ISO 12207 Process Alignment (informative)—Shows how the processes in each standard correspond in naming, numbering, and scope. The ISO 12207 processes provide specialized interpretations for software or address processes that apply only to software.
- Relationship to IEEE Standards (informative)—This standard was developed as a coordinated effort between IEEE and ISO. This annex provides a guide to related and supporting IEEE standards.

Guidance on the application and interpretation of ISO 15288 will be provided in a companion technical report ISO/IEC TR 24748, *Systems and Software engineering – Life cycle management – Guide for life cycle management* currently under development.

ISO 12207:2008, Systems and Software Engineering – Software Life Cycle Processes

ISO 12207:2008, *Systems and software engineering – Software life cycle processes* is a framework for addressing the processes, activities, and tasks used for the acquisition, development, and maintenance of software products. The standard is structured to recognize that the software life cycle processes are, in general, specializations of the system life cycle processes. Thus, ISO 12207 now corresponds very closely with ISO 15288.

The standard places software life cycle processes in the context of the organizational desire to define standard processes that fit the company's business needs. In other words, projects, where possible, will generally follow the processes defined by the organization rather than using processes designed to follow the ISO standard.

ISO 12207 is careful to distinguish between life cycle processes and life cycle models. No life cycle models are suggested by the standard. Rather, it is expected that the life cycle processes addressed by ISO 12207 will be mapped onto a life cycle model appropriate to the project. Ideally, the selected life cycle model will be one previously approved for use in the organization.

ISO 12207 groups the processes of the software life cycle into seven process groups, shown in Table 4.5 and Table 4.6. These processes are generally software-specific elaborations of the system life cycle processes identified in ISO 15288.

Table 4.5 ISO 12207:2008 System Context Processes

Group	*Process*	*Scope*
Agreement	Acquisition	Obtain products or services
	Supply	Provide products or services
Organizational Project-Enabling	Life Cycle Model Management	Maintain the organization's policies, life cycle models, processes, and procedures
	Infrastructure Management	Manage organizational infrastructure and resources to support projects
	Project Portfolio Management	Initiate projects needed to support organizational goals and policies
	Human Resource Management	Provide qualified, competent staff to projects
	Quality Management	Assure products and services meet quality goals and satisfy customers

(continued)

Table 4.5 ISO 12207:2008 System Context Processes (continued)

Group	*Process*	*Scope*
Project	Project Planning	Develop and disseminate project plans and schedules
	Project Assessment and Control	Determine project status and ensure project execution according to plans, schedules, and budgets
	Decision Management	Analyze and select decisions from among alternative choices
	Risk Management	Identify, assess, and manage risks
	Configuration Management	Maintain integrity of selected project and process outputs
	Information Management	Collect, maintain, and retrieve selected information
	Measurement	Collect, analyze, and report process and product measures
Technical	Stakeholder Requirements Definition	Identify system requirements for services needed by users and other stakeholders
	System Requirements Analysis	Transform stakeholder needs to technical system requirements
	System Architectural Design	Define the components and structure of a solution addressing the system requirements
	Implementation	Produce specified components that address architectural design requirements
	System Integration	Assemble system components in accordance with the architectural design
	System Qualification Testing	Confirm that design requirements have been met
	Software Installation	Install the software in the operational environment
	Software Acceptance Support	Support the acquiring organization in showing that the product meets requirements
	Software Operation	Use the software to provide specified services
	Software Maintenance	Sustain capability of the software through corrective, preventive, adaptive, and preventive actions
	Software Disposal	Retire and remove the software

Table 4.6 ISO 12207:2008 Software Specific Processes

Group	*Process*	*Scope*
Software Implementation	Software Implementation	Produce the software product or service
	Software Requirements Analysis	Define requirements for the system's software elements
	Software Architectural Design	Develop the top-level design for the software elements of the system
	Software Detailed Design	Create a design to describe coding and testing details
	Software Construction	Create executable software that implements the design
	Software Integration	Combine software units into integrated items in accordance with the software architectural design
	Software Qualification Testing	Confirm that software requirements have been met
Software Support	Software Documentation Management	Develop and maintain documentation during the software life cycle
	Software Configuration Management	Maintain integrity of selected software items
	Software Quality Assurance	Verify adherence to plans, standards, and procedures
	Software Verification	Confirm that software product and service requirements have been met
	Software Validation	Show that the software meets the stakeholder requirements for intended use
	Software Review	Maintain understanding among stakeholders of status and progress
	Software Audit	Independently verify compliance with requirements, designs, and plans
	Software Problem Resolution	Resolve all identified problems
Software Reuse	Domain Engineering Process	Develop and maintain reusable assets for specified domains
	Reuse Asset Management	Manage reusable assets throughout their life
	Reuse Program Management	Plan, manage, and control reuse opportunities

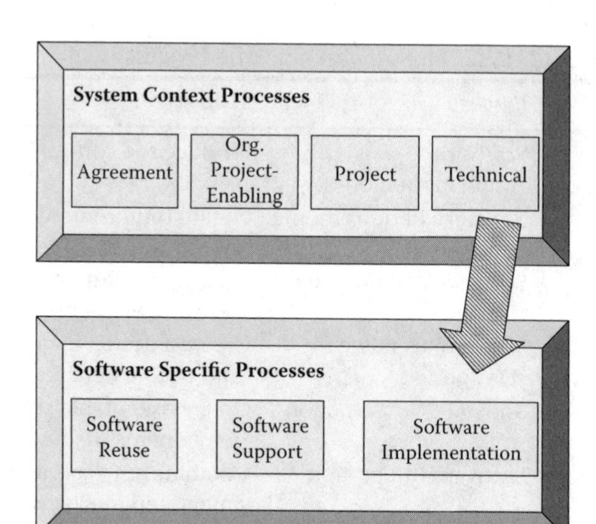

Figure 4.4 ISO 12207 life cycle processes.

The seven process groups may be organized into two sets. The first set, the System Context processes, shown in Table 4.5, are most useful for a stand-alone software system. The software life cycle processes in this set correspond very closely to the system life cycle processes defined in ISO 15288 and add software-specific details. This allows a project dealing with a software-only product or service to use a single standard, ISO 12207, without needing to use a second standard, ISO 15288, to provide the system context. The Software Specific processes, shown in Table 4.6, add the details needed for software implementation within a larger system.

These groupings are only a convenience because the life cycle processes are interrelated and rely on one another. Although these processes may be grouped and emphasized differently, depending on the point of view of a particular role, one way to look at them is shown in Figure 4.4.

As in ISO 15288, this standard defines the software life cycle processes by describing:

- The purpose of the process
- The desired outcomes resulting from process implementation
- The activities and tasks used to produce the desired outcomes

System Context Processes

The System Context processes shown in ISO 12207 use names and section numbers identical to those in ISO 15288 for the processes in the Agreement, Organizational

Project-Enabling, and Project processes. For each of these processes, ISO 12207 adds software-specific information to the more broadly defined processes in ISO 15288. To avoid needless repetition, we will not duplicate the discussion of those processes here.

For most processes in the Technical process group, ISO 12207 uses slightly different names. For example, "Architectural Design" and "Operation" in ISO 15288 become "System Architectural Design" and "Software Operation" in ISO 12207. The Technical processes of ISO 12207 contribute to satisfaction of the corresponding processes in ISO 15288 and contribute software-specific specializations to those processes. Except for the Implementation process, users may find it more convenient to follow ISO 15288 rather than the processes described in ISO 12207. The Implementation process is a placeholder process, pointing to the Software Implementation process for software-specific implementation activities.

Software Implementation Processes

This category provides the software-specific details for the Implementation process.

Software Implementation

This process is the software-specific specialization of the ISO 15288 Implementation process. Here, the appropriate life cycle model is selected and development activities are mapped to that model. Standards and procedures are then selected and tailored to fit the needs of the project and implementation plans are developed. The results of implementation planning are documented and maintained under configuration management.

The Software Implementation process is supported by six lower-level processes:

- Software Requirements Analysis
- Software Architectural Design
- Software Detailed Design
- Software Construction
- Software Integration
- Software Qualification Testing

Software Requirements Analysis

In the Software Requirements Analysis process, the system requirements allocated to each software item are analyzed to develop the software requirements. Software requirements must be traceable to, and consistent with, the system requirements and system architecture. Software requirements include specification of functional requirements; performance requirements; interfaces; safety and security requirements; qualification and acceptance requirements; and user documentation, operations, and maintenance requirements.

Software Architectural Design

The Software Architectural Design process translates the software requirements into a high-level design, allowing the requirements to be allocated to software components. The interfaces between those components are defined and the high-level design, interfaces, database design, and test requirements are all documented. Preliminary user documentation and software integration schedules are also developed as part of this process.

Software Detailed Design

The Software Detailed Design process develops the design of each software component to finer levels to allow them to be coded and tested. Interface definitions, test requirements, and integration and test schedules are refined. The detailed design, database design, and interfaces are documented.

Software Construction

Using the detailed design specifications, each software unit and database is coded and tested. The software integration schedule is updated and documentation is revised as needed.

Software Integration

In the Software Integration process, a documented plan for the integration of software units is developed and executed. The results of integration and testing are documented to ensure that the software requirements have been satisfied and that the software components are ready for software qualification testing.

Software Qualification Testing

The Software Qualification Testing process demonstrates that the software product performs as specified. Test plans and results are documented. Audits of the software qualification results may be conducted as part of the audits of system qualification testing.

Software Support Processes

The Software Support processes are primarily used to assist the Software Implementation processes, but may also be used with other processes. In some cases, they also provide software-specific specializations for the processes found in the System Context grouping. They are:

- Software Documentation Management (supports the Information Management process)
- Software Configuration Management (supports the Configuration Management process)
- Software Quality Assurance
- Software Verification
- Software Validation
- Software Review
- Software Audit
- Software Problem Resolution

Software Documentation Management

The Software Documentation Management process captures the results of implementing software life cycle processes. A plan identifying the documents to be developed specifies documentation standards and the schedule, methods, and responsibilities for development, review, control, and production of those documents.

Software Configuration Management

Software items are identified and controlled through the Software Configuration Management process, which is used to ensure the completeness and consistency of software items.

The process begins with the development of a plan for configuration management, which includes responsibilities and the procedures for performing configuration management activities. ISO 12207 takes a fairly standard view of the software configuration management process as comprising five major activities:

- Configuration identification—Specifying the software items and versions to be controlled
- Configuration control—Recording, reviewing, approving, and implementing requests to modify software items
- Configuration status accounting—Monitoring the status and history of software maintained under configuration management

- Configuration evaluation—Ensuring the functional and physical completeness of software items
- Release management and delivery—Controlling the release and delivery of software and documentation

Software Quality Assurance

The Software Quality Assurance process is used to assure that software products conform to their standards and that the processes used follow their plans. To provide that assurance, objective evaluations are necessary. In addition to having appropriate authority and resources, those who perform the Software Quality Assurance process are expected to have organizational freedom from those developing the product or delivering a service.

Plans for software quality assurance include identification of quality assurance procedures, responsibilities, and schedules. Software quality assurance activities are expected to be coordinated with other supporting processes such as Software Verification, Software Validation, Software Review, Software Audit, and Software Problem Resolution. Product assurance requirements include ensuring that plans required under the contract are developed and that products satisfy their requirements. Process assurance requirements address the need for those who are implementing software processes to follow plans, measure processes and products, and properly train the project staff.

Software Verification

The Software Verification process is used to determine if products meet their requirements. The requirements may be part of a customer specification or may come from a prior life cycle process such as Software Detailed Design or Software Construction. Unlike the Software Quality Assurance process, the Software Verification process is not necessarily implemented by those independent of the developers or operators, although the use of independent third parties may sometimes be appropriate. Problems detected by the Verification process are handled through the Software Problem Resolution process.

Verification planning includes determination of the need for verification, evaluation of the criticality of verifying specific products, and selection of verification approaches. Verification activities include the following:

- Requirements verification
- Design verification
- Code verification
- Integration verification
- Documentation verification

Software Validation

The Software Validation process is used to determine if the completed software system supports its intended use. As defined by ISO 12207, software validation is clearly focused on testing the end product and not on intermediate work products created during the product's life cycle. Validation testing includes stress testing, testing at boundaries, testing the ability of users to conduct their intended tasks, and testing the software in its target environment. Validation issues are resolved through the Software Problem Resolution process.

Software Review

The Software Review process is used to help stakeholders maintain a shared view of project management and technical status. Project management reviews focus on evaluating progress against plans, monitoring resource allocation, and managing risks. Technical reviews are used to evaluate software products and services to confirm that they meet their specifications, are developed according to plan, and are ready to progress to the next scheduled activities. The Software Problem Resolution process is used for problems found by the Software Review process.

Software Audit

The Software Audit process is used for independent evaluation of compliance with requirements, specifications, contractual agreements, and plans. Problems and discrepancies discovered through the Software Audit process are handled by the Software Problem Resolution process.

Software Problem Resolution

As a supporting process, the Software Problem Resolution process is used to address problems discovered during the execution of all other software life cycle processes. When problems are detected, appropriate parties are notified, the problems are analyzed, dispositions are determined, and status is monitored and reported. Records of problems and resolutions are maintained.

Software Reuse Processes

This group contains processes to help an organization take advantage of knowledge and work products developed across multiple projects. The Software Reuse processes are:

- Domain Engineering
- Reuse Asset Management
- Reuse Program Management

Domain Engineering

The Domain Engineering process is used to capture and reuse architectures, models, concepts, and assets within the scope of particular domains. Using a domain engineering plan, the domain specialists identify the anticipated needs of stakeholders, construct models, and conduct domain reviews. Architectures and other assets are acquired or developed and are maintained.

Reuse Asset Management

The Reuse Asset Management process manages assets designated for reuse throughout their life. Facilities for asset storage and access are implemented. Potentially reusable assets are reviewed, evaluated, and classified. If accepted, the asset is maintained under configuration management and made available to the organization.

Reuse Program Management

The Reuse Program Management process is used to provide the overall management of an organization's reuse program so that opportunities for reuse may be effectively executed. As part of this process, domains suitable for reuse exploitation are identified, the ability of the organization to take advantage of its reusable assets is evaluated, and recommendations for improvements to the organization's reuse capabilities are identified. The reuse program is periodically planned, monitored, and reviewed.

Annexes

The ISO 12207 standard includes a number of annexes, both normative and informative. The topics addressed include:

- Tailoring Process (normative)—Summarizes the activities for tailoring the standard to meet the needs of an organization or project.
- Process Reference Model for Assessment (normative)—Process assessments under ISO 15504 require the use of a Process Reference Model (PRM). ISO 12207 satisfies the requirements of a PRM. This annex provides considerably more detail than the corresponding informative section in ISO 15288.

- History and Rationale (informative)—Summarizes the history of the standard, the goals of the harmonization effort, and the process definition constructs shared with ISO 15288.
- ISO 15288 and ISO 12207 Process Alignment (informative)—Same as the corresponding informative section in ISO 15288.
- Process Views (informative)—Similar to the corresponding informative section in ISO 15288.
- Example Process Descriptions (informative)—Outlines some additional processes that might be added to organizational standards.
- Relationship to IEEE Standards (informative)—Similar to ISO 15288, this annex provides a guide to related and supporting IEEE standards.

ISO 20000:2005, Information Technology – Service Management

ISO 20000:2005, *Information technology – Service management* (ISO 2005a, 2005b) is used to provide an integrated process approach to the delivery of managed information technology services. ISO 20000 is largely based on the BS 15000 standard, which, in turn, is closely related to the Information Technology Infrastructure Library (ITIL). However, the ISO 20000 and BS 15000 frameworks differ from ITIL in their intent, structure, style, and detail. This is not to say that they are not synergistic; an organization that has adopted ITIL will be better positioned to achieve ISO registration than the one that has not. In other words, the principles presented in ISO 20000 can be successfully implemented with ITIL. The ISO 20000 standard limits its purpose to the provision of information technology (IT) services, although many of its concepts may be extended and interpreted for other services, such as systems integration and personal services. Development of applications or computer-based products is not addressed in this standard. The salient features of ISO 20000 are described later.

IT service providers can no longer limit their product to technology alone; they must ensure that they provide value-added services and foster mutually profitable relationships with their customers. Service Management (SM), defined in ISO 20000 as "management of services to meet the business requirements," establishes the concept that business requirements are the main driver for demanding the processes needed for successful implementation of service management.

ISO 20000 and ISO 9001 are both based on an integrated process approach where it is recognized that an enterprise has many functions and processes that have to operate and be managed as a system, rather than as independent activities. ISO 20000 helps its users identify those functions, and manage, control, and improve them. By doing so, an organization is poised to manage those functions more efficiently and effectively. In addition, ISO 20000 and ISO 9001 have many processes in common. As we will discuss in chapter 6, this commonality can be successfully exploited when developing an overall process approach and architecture.

ISO 20000, one of the first standards for information technology service management (ITSM), has two parts—Part 1: Specification and Part 2: Code of practice. Part 1 lists requirements, and Part 2 provides guidance, recommendations, and best practices for implementing those requirements. Part 1 is very compact; the requirements are listed on only 12 pages using "shall" statements. Part 2, at 31 pages long, is a little more descriptive, repeating the requirements of Part 1 and providing explanation, elaboration, and necessary documentation for each of those requirements. Part 2 uses the "should" construct in most of its statements. Unlike some of the ISO standards described earlier in this chapter, it is mostly self-sufficient, having very few cross-references. To understand this standard, both parts should be studied and implemented together, but no other ISO standards are required. Although ITIL is not referenced by ISO 20000, many organizations make use of the extensive descriptions provided by ITIL when implementing the ISO-required processes.

The ISO 20000 structure is shown in Figure 4.5. Sections 1 and 2 address the scope and foundation definitions. Sections 3, 4, and 5 provide the infrastructure for performing other requirements of the standard. Those three sections align very well with the requirements of ISO 9001 and address management responsibility, documentation requirements, training, and planning and implementation of service management. Sections 6 through 10 address specific requirements for successfully delivering and improving services. ISO 20000 clearly states that it is not intended for product assessment but both parts of the standard may be used to support development of service management best practices.

ITSM Infrastructure

As noted, ISO 20000 sections 3, 4, and 5 describe the requirements for an infrastructure that will support service management. The infrastructure covers several major topics: management responsibility, training, documentation, and requirements for planning and implementing service management and changes to service management. To be successful, senior management should:

- Establish service management policies, objectives, and plans
- Communicate their importance
- Establish responsibility for SM

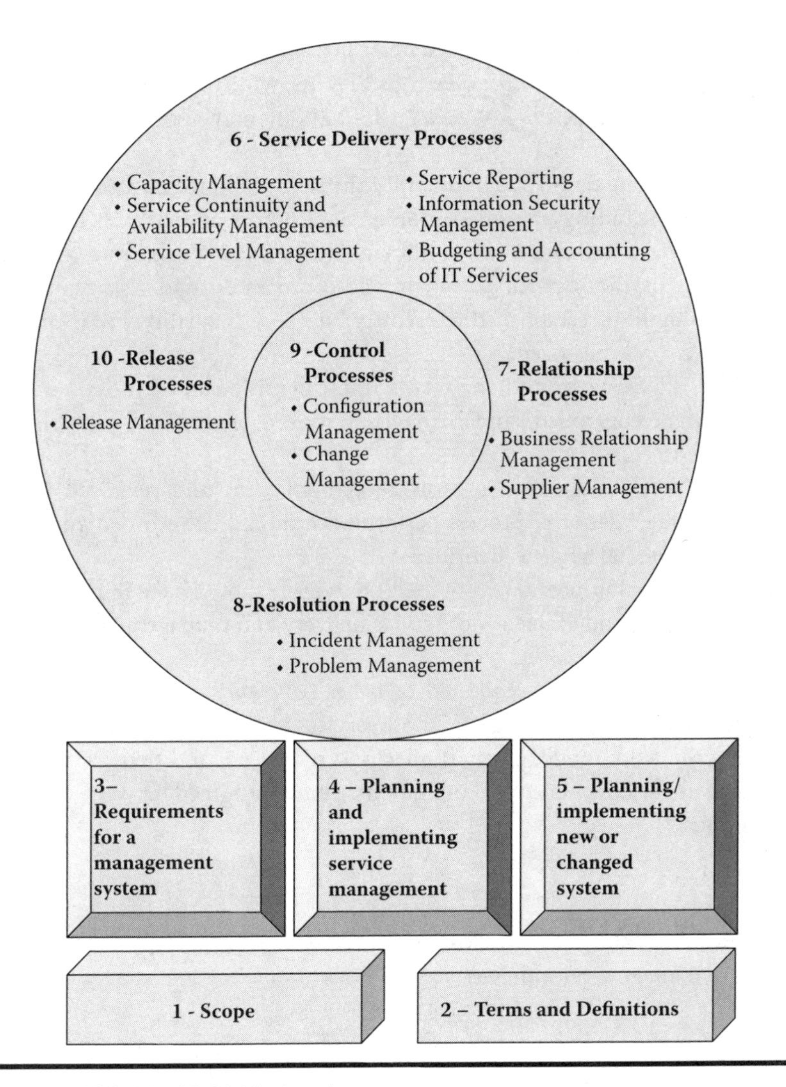

Figure 4.5 ISO 20000:2005 structure.

- Provide necessary resources
- Establish a training capability to ensure that appropriate skills are available when needed
- Ensure that necessary documentation is provided that describes its policies, processes, procedures, plans, and service level agreements (SLAs)
- Ensure that the records showing implementation of policies, procedures, and plans are collected

Like ISO 9001, the planning described in ISO 20000 requires the use of the Shewhart cycle (Plan–Do– Check–Act, or PDCA), which was discussed in chapter 2. Implementation of the PDCA cycle for ISO 20000 may be viewed as follows:

Plan—The organization should identify the scope, objectives, and requirements for SM, including the quality of service, processes that will be followed, resource requirements (such as human, facilities, and tools), roles and responsibilities of those participating in those processes, and risk management. Plans should be established to identify how service quality will be managed, assessed, and improved.

Do—The IT services, and the teams implementing those services, should be measured, controlled, and managed. Progress against the plan should be reported.

Check—Processes should be measured, monitored, and reviewed (audited). Objective evidence of process performance should be recorded and any necessary remedial actions identified.

Act—Continual improvement in the PDCA cycle should be used to improve the effectiveness and efficiency of service delivery and management.

When a new service is requested or an existing service is changed, management is responsible for ensuring that those services are performed according to their plans; that the work is adequately funded and resourced; and that service delivery is managed at the cost, schedule, and quality mutually agreed between the supplier and customer.

Service Delivery Process

In order to manage service delivery:

- Each service must be defined
- Agreements should be documented (using a service level agreement, or SLA)
- SLAs need to be agreed to by all relevant stakeholders
- Services should be controlled by a configuration management process
- Services should be reviewed and revised at regular intervals
- Services must be monitored against established targets

The SLA is an essential requirement for the provision or receipt of services and is central to the requirements of this standard. The SLA is a written agreement between the customer and service provider that defines the services to be provided and the service levels. The objective of the service delivery process is to provide the defined services without disruption while exhibiting the desired availability in all agreed upon circumstances. Delivered services must be monitored against the service level targets, trend information should be tracked, satisfaction with the service analyzed, and noncompliances and issues collected and resolved. Requirements for service availability (the ability of a component or service to perform its required function at a stated instant or over a stated period of time) and continuity (the ability to recover and restore the IT service infrastructure) must be defined in the SLA. Those requirements should be evaluated periodically (for example, yearly) and whenever changes (such as environmental conditions) have been introduced that may affect either availability or continuity. The service continuity plan should contain the steps needed to return to normal operation after the occurrence of a major event that prevented continual operation.

Two additional processes have to be addressed: capacity management and information security management. Capacity management must be planned, based on the current and predicted capacity and performance requirements, required workload, resource utilization, and network throughput. Capacity management plans also need to address potential internal and external change requirements and their impacts, including the costs of service upgrades. Capacity plans should be monitored and any departure from plans should be addressed to ensure adequate performance and capacity. Information security should be managed within all service activities. ISO 20000 recommends using the ISO/IEC 17799 (ISO 2005c) standard to implement information security policy requirements, document security controls, and report and record security incidents.

Relationship Processes

Relationships are managed through two related processes: Business Relationship Management and Supplier Management. The purpose of the Business Relationship Management process is to establish and maintain mutually favorable relationships between service providers and customers, whereas the purpose of the Supplier Management process is to ensure effective provision of quality service from the service provider's suppliers.

The Business Relationship Management process is based on an understanding of customer requirements. To gain this understanding, both the service provider and the customer have to communicate, for example, via meetings and reviews. All stakeholders, such as subcontractors, user groups, and operation and maintenance staff, should participate in those meetings where changes to the scope of services, SLAs, contracts, and business needs are reviewed. As part of the Business Relationship Management process, two related processes have to be established, namely, a *complaint process* and a *customer feedback process*. The complaint process enables complaints to be collected, recorded, investigated, acted upon, reported, and finally closed. The customer feedback process enables collection and analysis of customer satisfaction surveys and measurements. The survey analysis can then be used to identify processes that need improvement to alleviate customer dissatisfaction. Due to the very visible position of this process, the standard requires appointment of an individual responsible for customer satisfaction and the whole business relationship process. Collected complaints as well as the lessons learned from managing business relations may be input to process improvement plans.

The Supplier Management process addresses the relationships between the service provider and its suppliers and subcontractors. That relationship is similar and parallel to the service provider's relationships with its customers. There should be an SLA between the service provider and its suppliers that covers topics similar to those in the service provider's SLA with its customer. Moreover, those two SLAs should be aligned for the supplier to satisfy the service provider's requirements and, in turn, its business requirements. The interfaces between the provider's and supplier's processes should be documented, including their roles and responsibilities. A review process should be in place for reviewing SLAs, contracts, changes to SLAs and contracts, and disputes. Any changes to formal agreements are subject to further consideration under the Change Management and Change Control processes. The performance of suppliers should be monitored and lessons learned should be collected for potential inclusion in the service improvement plan. The service provider should also develop a process for addressing the expected end of service, early end of service, and transfer of service to another party.

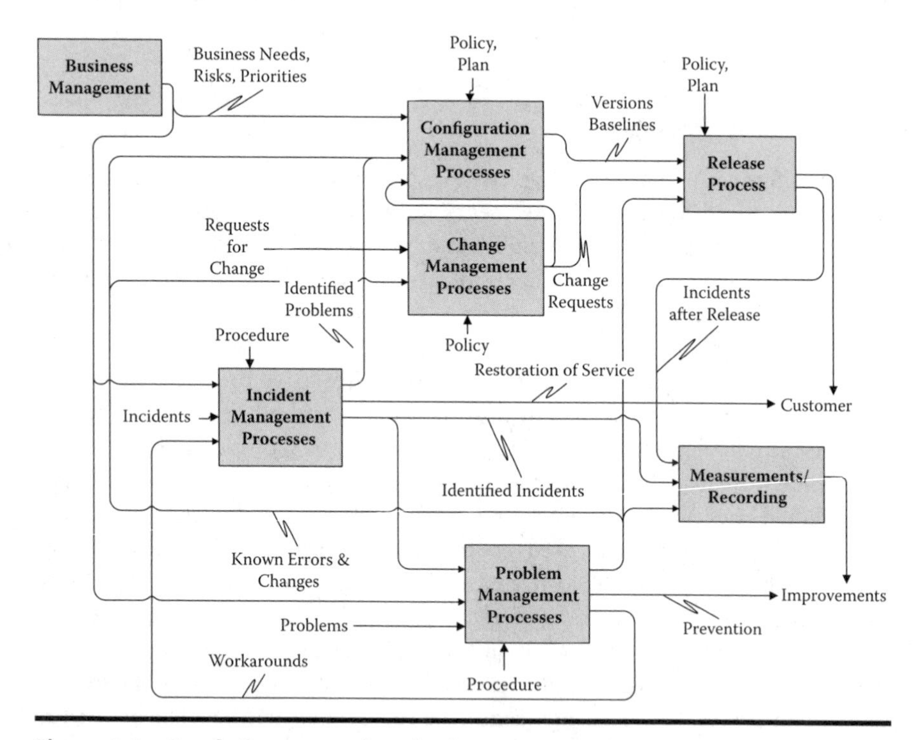

Figure 4.6 Resolution, control, and release processes.

Resolution, Control, and Release Processes

Using the IDEF0 notation explained in chapter 3, Figure 4.6 shows interactions among the resolution, control, and release processes. The Measurement and Recording process, which is required for collecting and distributing measurements among all processes, is also shown. The purpose of the Resolution, Control, and Release process groups is to provide incident-free service to the customer. In addition, the service provider must be able to take into account those changes that are necessary for successful operation of the customer's enterprise and have processes in place that will analyze problems and incidents, develop workarounds, restore service if changed applications are not properly corrected, and ultimately improve its processes. We will briefly discuss each of those processes and their relationships.

Resolution Processes

Resolution processes are divided into the Incident Management and Problem Management processes. Those two processes are distinct but closely related. The purpose of the Incident Management process is to promptly restore required service, and the purpose of the Problem Management processes is to proactively identify and analyze causes of the incidents and manage them to closure.

Through the Incident Management process, incidents are reported, collected, identified, classified as to their severity, and tracked to closure. The priority of addressing the incident is usually established by service supplier management and agreed to by the customer. It is based, for example, on the incident's business impact, risk of the loss of service, or the need for the continuation of operations. Because incidents cannot always be immediately resolved, the Incident Management process should take into account the possibility of operation with degraded service and provision of workarounds until full service is restored.

The Problem Management process addresses the analysis of incidents, determination of their root causes, and proactive prevention of recurrence of the incidents and known errors. Known errors are problems for which the root cause and resolution method is known. Such information must be made available to the Incident Management process. The Problem Management process typically describes steps for the identification of problems, minimization or avoidance of their impact, recording, classification, updating escalation, resolution, and closure. The Change Management process addresses changes required to remedy identified problems.

Throughout the execution of the resolution processes, the customer should be kept apprised of progress. By trending the incidents, their number, severity, time to restore service, and so on, the service supplier will be able to develop preventive and improvement actions.

Control Processes

The Control processes are the Configuration Management and Change Management processes. Those two processes are very closely related and, in practice, they should be integrated. The purpose of the Configuration Management process is to identify and control components of the service and infrastructure and ensure their integrity while the purpose of the Change Management process is to assess all requested changes and then implement only those changes that have been reviewed and approved by the service stakeholders.

The Configuration Management process provides criteria for selecting service or infrastructure components to be controlled based on their criticality, business needs, or risk of failure. It enables their identification, description of their relationships and interfaces, and capture of the necessary documentation. Versions of the identified components should be controlled, tracked, and recorded in a database and will form a baseline to be used in the service release. Requested changes to the identified components should be traceable and auditable, and the impact of the change should be assessed based on its importance to the service delivery. Approved changes are provided to the Change Management process for further analysis.

The Change Management process enables change requests to be recorded and classified, and their risks, benefits, and impacts on business assessed. The change management concepts described in ISO 20000 are different from the concepts of change management commonly defined in business process reengineering (BPR), which describe a formal approach for transitioning from a current state to a desired state. When changes are approved and implemented, they are reviewed to determine success or failure, and appropriate actions are documented. A special process should be established for addressing high priority or emergency changes. The Change Management process must also provide for reversing or remedying the changes when the implementation was not successful. By analyzing change records and data for trends, the service provider will be able to develop improvements that may help eliminate their causes and improve service.

Release Processes

The Release process is closely related to the Control processes. Release plans and schedules should be developed, reviewed by the service provider's management, and approved by customers, end users, operation and support staff, and other stakeholders. Information contained in the release baselines should be coordinated through the Configuration Management and Change Management processes, as well as through the Incident and Problem Management processes. In addition, the Release Management processes interact with the installation, handling, packaging, and delivery processes. There should be provisions for emergency releases in cases where the resolution of an incident is critical to the customer's operation. By measuring success and failure of releases, the service provider will be able to develop plans for improving service and better estimate required resources.

Summary

In this chapter, we have summarized the content of five ISO standards. One, ISO 9001, is suitable for general application across organizations of all types in many different industries. The others are directed toward application in specific areas, namely, systems engineering, software engineering, and information technology services.

All of these standards share a process approach, rather than a compliance approach. That is, there is a recognition that the activities that transform inputs to outputs (i.e., processes) may be grouped and sequenced in different ways and that processes interact with each other. Processes need to be documented, their performance needs to be measured, and they should be improved over time.

Processes may be grouped and categorized differently in each standard, even where similar topics are being addressed. Several standards rely on cross-references to other standards to supply additional discussion and definition. Unfortunately, standards are revised and amended asynchronously and the revision process takes many years. Thus, we often find inconsistencies, such as references to obsolete standards. Efforts at harmonization of ISO standards (and harmonization of ISO standards with others, such as IEEE standards) have been underway for several years and are continuing.

Chapter 5

Framework Mapping

In this chapter, we will show how to use maps between several frameworks. We will first show the maps relating CMMI v1.2 to each of the selected ISO standards. Later, based on the information obtained from those maps, we will show how to use them and how to detect framework synergies and gaps. Keep in mind that maps between two frameworks are great tools, but they are only tools. An understanding of the frameworks is needed to use them effectively.

The Mapping Process

One problem that arises when developing maps between two frameworks is that the authors of each framework are not bound to follow the approaches taken by other frameworks. This often results in incompatible vocabularies, differing architectures, dissimilar process decompositions, and many-to-many relationships among framework requirements. To illustrate these complexities, consider the map between ISO 20000 and CMMI v1.2 shown in Figure 5.1. In this figure, we can see that the requirements of ISO 20000 paragraph 4.4.3 "Continual improvement activities" map to several practices in the OPF, OPD, MA, and OID PAs. Examining one of those CMMI practices, we can also see, for example, that OPD SP 1.1 Establish and Maintain the Organization's Set of Standard Processes, maps to several requirements spread across the ISO 20000 framework. Simple one-to-one relationships are not always to be found.

This many-to-many relationship is to be expected, because both frameworks are process-based and their decomposition into consistent components is not unique. Those components interact but their interaction is not always strong, so one must decide when to continue with the interaction chain. We will show in chapter 6 that

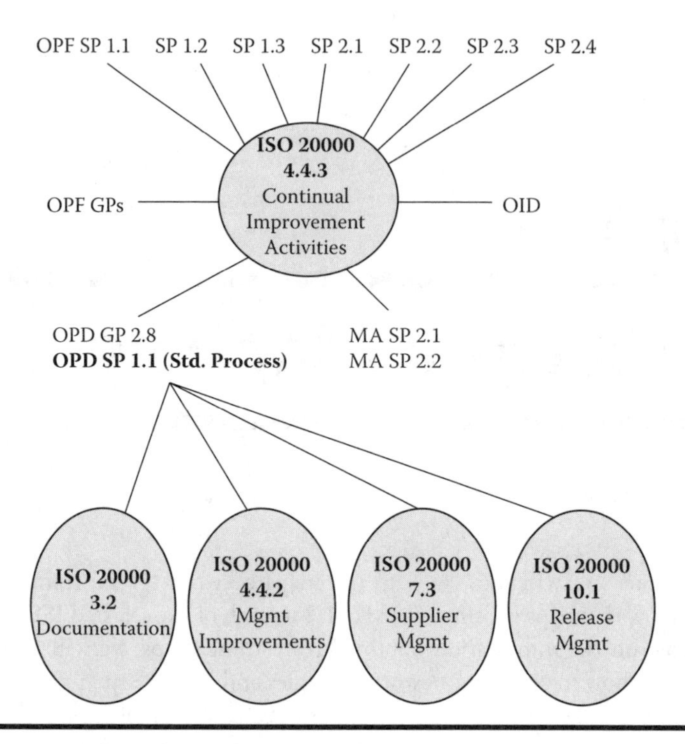

Figure 5.1 Mapping relationships example.

such interactions have a strong impact on the process architecture, creating numerous interfaces among process elements. Figure 5.1 also demonstrates that a map is just a tool that can be used to help develop an understanding of a lesser known framework by comparing it to a more familiar framework. A good understanding of each framework is mandatory before the maps can be effectively and efficiently used. Based on this brief discussion, we hope that the impossibility of developing a process map that is both accurate and objective can be appreciated. Nevertheless, the maps can be used for both process improvement and appraisals, as we will show in chapters 6 and 7, respectively.

Over the years, we have developed a process for mapping frameworks, which has been fine-tuned through the feedback we have received from users and reviewers. This process is shown in Figure 5.2.

First, we study the selected framework or standard to understand its features and consult the literature describing the framework and its use. For the ISO standards addressed in this book, there are additional ISO guidelines that should be considered when performing the mapping (ISO 2000c, 2005b, 2008c).

We have also established a set of ground rules that are related to framework terminology. In essence, we create a dictionary of terms that relate two frameworks. When there is agreement on the terminology used, the mapping can be

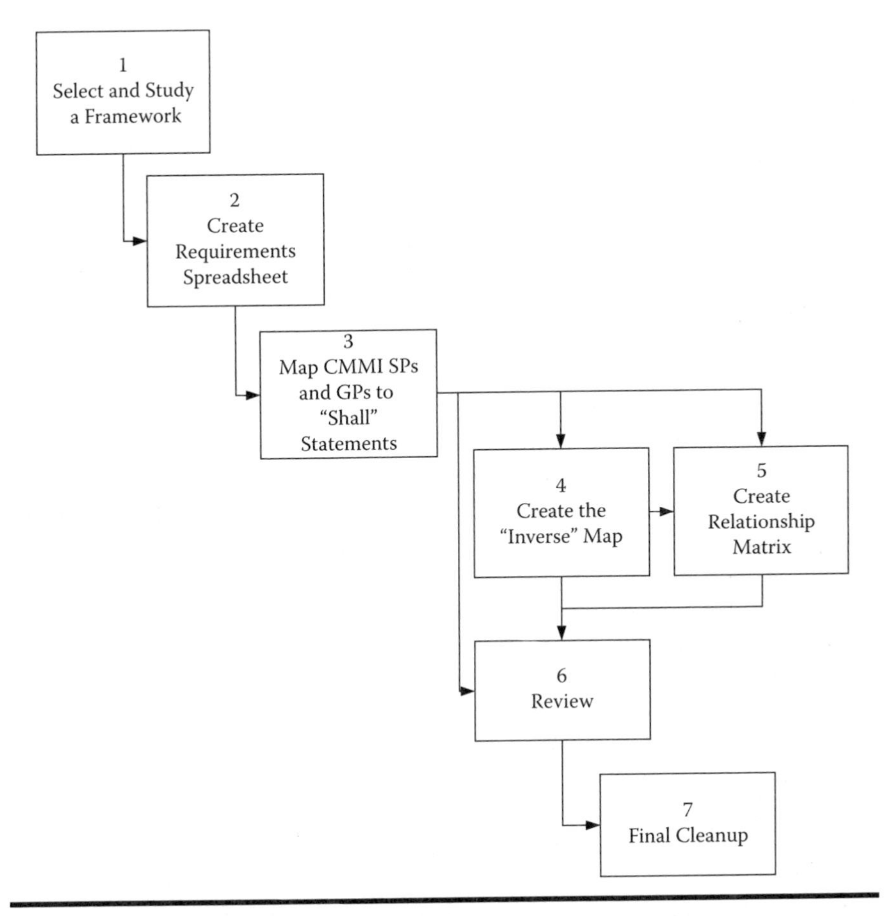

Figure 5.2 Mapping process.

more readily performed and accepted. In the sections that follow, we will provide those ground rules and associated dictionaries, where applicable, before we present the maps themselves.

Determining the appropriate granularity of maps between models is subjective. A map at a high level, for example, relating major sections of an ISO standard to entire CMMI PAs, may not provide enough insight into similarities and differences. A map at a very low level, on the other hand, results in an overwhelming number of matches, which also fail to properly illuminate framework relationships. Very low-level maps may show several dozen relationships without regard to their strength, making the true relationships difficult to discern. We have found that mapping at the "shall" level or, in some cases, at the level of each item in an enumerated list makes the most sense. We then capture the appropriate framework statement in a spreadsheet.

When a sufficient understanding of the framework has been reached, we start the actual mapping process, relating each "shall" or list item statement to the appropriate CMMI SP or GP. CMMI goals are, from an appraisal point of view, the required model components that describe the characteristics of PAs or their institutionalization. Each goal has other associated model components, SPs and GPs, which are expected for goal satisfaction. Thus, we map to SPs and GPs but never to CMMI goals. We then add columns to identify the PAs, SPs, GPs, confidence factors (discussed later), and capture additional comments. We often use a shortcut notation when, for example, an ISO shall statement maps to a particular GP in all PAs. In that case we enter "All PAs" in the PA column and the relevant GP in the column for CMMI Practices.

This map from the source framework to CMMI is the "direct" map. As shown in Figure 5.2, after creating a direct map between each ISO standard and CMMI and entering it into a commercial tool,* we create an "inverse" map. The inverse maps provide a consistency check for the mapping process. During the direct mapping process, from the source framework to CMMI, there are times when there is a degree of uncertainty that the best fit has been found. The inverse maps help provide insight into resolving that uncertainty. In addition, as we will see in chapters 6 and 7, the inverse maps can be used to guide process improvement, whereas the direct maps are better suited to determining compliance.

We characterize each map as strong, medium, or weak. For that purpose we use a confidence factor, shown in Table 5.1, to describe the strength of the mapping.

The *strong* and *no map* definitions are self explanatory, but the *weak* and *medium* confidence factors are more subjective. Because of the particular characteristics and domain coverage of the ISO standards, a perfect match between their requirements and CMMI is seldom possible, but, with proper interpretation, one can see the emerging relationships that can be used as a basis for process improvement. These interpretations need to be used with caution when performing appraisals. A weak map will require a major effort to ensure compliance when implementing the framework, whereas a medium map will, in most cases, only require interpretation. For example, CMMI requires projects to develop a project plan (PP SP 2.7). If that plan addresses service management, it would certainly satisfy the ISO 20000 requirement for service management planning. In contrast, the complaint process required by ISO 20000 in paragraph 7.2 "Business relationship management" has no equivalent CMMI practice and requires the development and implementation of such a process.

Since most maps show many-to-many relationships, there are numerous instances where several CMMI practices will be mapped to a single ISO requirement and vice versa. Not all of the relationships will be of the same strength; some may be strong and some may be weak. With a confidence factor in place, it becomes

* The tool used is Integrated Systems Diagnostics (ISD) Appraisal Wizard, which contains Model Wizard and Model Mapper. Most of our work has used Model Mapper.

Table 5.1 Mapping Confidence Factors

Mapping Confidence	*Description*	*Comment*
0	No map	
30	Weak	The statement in the ISO standard does not clearly correspond to any CMMI practices, but the ISO statement can be interpreted and implemented using CMMI process area components.
60	Medium	The match is not complete, but with some interpretation, CMMI may satisfy the ISO requirement.
100	Strong	There is a strong relationship between the ISO requirement and the CMMI practice.

evident which maps are most important for process improvement and users of the maps may, for example, decide to keep only those that show strong relationships. As we will see in chapter 7, even a weak map that leads to the implementation of a required process may still produce evidence that can be successfully used in an appraisal.

We have tried to take a somewhat strict stance when assigning confidence factors and have been parsimonious in providing 100 percent confidence factors where the relationship between the framework elements is not sufficiently strong. Stretching the scoring of mapping strength is counterproductive and misleading. We prefer to use comments and describe those maps where multiple interpretations are possible, particularly where the maps are weak.

We have established additional mapping ground rules, which become integral parts of each specific mapping. For example, in ISO 9001, a typical shall statement may include several enumerated items. Do we map each item or the entire shall statement? The answer, as always, is "It depends." This is another area where subjectivity in mapping comes into play. In some instances, where the enumerated items describe features of the shall statement, it is sufficient to provide a single map to the whole statement. In cases where the enumerated items outline detailed requirements, it is more correct and useful to provide a mapping for each individual item. The method selected is clear from the context of the mapping.

A significant concern when performing mapping is maintaining consistency in the mapping approach. For example, if an ISO requirement maps to an instance of CMMI GP 3.1, does it also map to GP 2.2? Or, if the ISO requirement maps to GP 2.2, does it also map to GP 2.3, GP 2.4, and so on? Similarly, how should the GP–PA relationships (shown in Table 7.2 in Chrissis et al. 2006) be handled? We will explain this in some detail in chapter 6. Throughout the mapping process, care must be taken to address the nuances and the approach to decomposition of requirements taken by each framework. For example, ISO 9001 section 4.1 "General requirements" requires an organization to perform a number of activities such as identifying processes, determining their sequence, and providing resources. This top-level requirement is distinct from ensuring that planning was done (paragraph 5.4.2) or performing the planning for product realization itself (paragraph 7.1). Hence, mappings for top-level requirements will not generally include the same level of detail to be found in lower-level requirements.

The spreadsheet thus shows the ISO requirement, CMMI PA, CMMI practice (SP or GP), confidence factor, and comments. The mappings discussed in this book were developed followed the approach described here and may be found in the appendices.

In most cases, we have developed relationship matrices that relate one framework to another in the form shown in Figure 5.3.* When generating the relationship matrix, individual shall-statement maps are rolled up to a major heading or requirement so relationships may be seen and are not lost in a fog of details. The detailed shall-level maps are always preserved in the spreadsheets and stored in the tool.

We found such matrices very attractive for both understanding and presenting framework relationships. As shown in Figure 5.3, we usually indicate CMMI practices in the vertical dimension and the source framework in the horizontal dimension. An *x* in a cell indicates a relationship. In cases where an ISO requirement is supported by many enumerated items, we sum the hits for each individual ISO section or subsection. Thus, some maps, such as the ISO 20000 mapping to CMMI, will contain numbers instead of *x* in their cells. In addition to just indicating that relationships exist, we sum the number of hits in each column and row. Those sums then indicate, for example, those requirements that don't have a map and those that have a large number of hits, and thereby help us to detect mapping patterns that may reveal relationships between large numbers of requirements. In cases where an ISO requirement is supported by many enumerated items, we sum the hits for each individual ISO section or subsection. When analyzing the maps, those totals facilitate analysis and help in obtaining an overall picture of the relationship.

The review, shown in Figure 5.2, compares the direct and inverse maps and identifies potential inconsistencies. In addition, we solicit comments from colleagues who are familiar with those frameworks, consult published maps where appropriate, and adjust the maps accordingly.

Since the ISO standards are protected by copyright, we cannot list the shall statements verbatim when we prepare the maps for publication. We do, however, provide paragraph numbers and keywords to allow users of the standards to easily determine the relevant requirement and context.

* The mapping relationship was first introduced to us by Lockheed Martin and has been published on the SEI Web site. Another example is found in the maps developed at Griffith University in Australia.

ISO xxxx:yyyy CMMI	Section 4	Subsection 4.1	Shall stmnt# 1	Shall stmnt# 2	Shall stmnt# 3	Shall stmnt# 4	Shall stmnt# 5		Section 5	Subsection 5.1	Shall stmnt# 1	Shall stmnt# 2	Shall stmnt# 3	Shall stmnt# 4	Shall stmnt# 5
PA x															
SP 1.1			x								x				x
SP 1.2					x										
GP 2.1						x					x		x		
GP 2.2						x					x				
PA y															
SP 1.1			x		x	x	x							x	
SP 1.2											x				
GP 2.1															
GP 2.2				x										x	

Figure 5.3 Mapping relationships in a matrix form.

As a basis for discussion, we will use our correlation matrices as road maps for presenting the total mapping picture. The matrices can be easily manipulated while still preserving and illuminating framework relationships that would be otherwise difficult to see. However, they can grow quite large, so we have disaggregated them for presentation in this book. For example, we have split the overall matrix into one that addresses only the SPs and another one that addresses only the GPs. Then we further split the SP matrix into four groups based on categories, similar to the CMMI continuous representation. These disaggregated matrices helped us to better understand relationships between ISO standards and CMMI. We will use the disaggregated matrices in our mapping examples in this book. For clarity in the tables that show mapping relationships, we compress those ISO requirements that did not generate any maps to CMMI. This style is particularly important when only one or two of several enumerated requirements are mapped. For example, if a shall statement has enumerated requirements a through h, and only requirements d and h are mapped, then the unmapped requirements (a, b, c, e, f, and g) will not be shown in the matrix.

In this chapter we present maps between the ISO source frameworks and the CMMI target. The real value is in interpreting what those maps are telling us about the strengths and weaknesses of each framework with respect to CMMI and the relationships among the frameworks. The matrices, which help us visualize mapping in both directions, from ISO to CMMI and from CMMI to ISO, are important because:

- There are ISO registered organizations that would like to move toward CMMI maturity levels and vice versa.
- There are organizations that are primarily interested in the process improvement impact of using multiple frameworks.

By considering bidirectional mappings we can learn more about each individual framework and its potential for supporting process improvement.

In the following sections we discuss each individual ISO-to-CMMI map, the approach taken, and draw some conclusions regarding framework synergy. In chapter 6 we will take a closer look at several maps and discuss how those maps can be used for creating a successful process improvement strategy with particular emphasis on creating a versatile process architecture. In chapter 7 we will address the appraisal issues stemming from those maps.

The maps between the ISO standards and CMMI that are discussed in this chapter may be found in the appendices.

ISO 9001:2000 to CMMI Maps

We indicated earlier in this chapter that having a terminology comparison between the frameworks helps the creation and understanding of the map. Terms that are essentially equivalent in both standards (such as *system* or *process*) are not discussed. Table 5.2 shows how some major ISO and CMMI terms are compared (Mutafelija and Stromberg 2003a, 2003b; Stromberg and Mutafelija 2002, 2004).

By inspecting the ISO 9001–CMMI relationship matrix, we realize that in addition to obvious individual mapping strengths and weaknesses, the more global aspects of those relationships will have to be addressed. An examination of that matrix reveals several trends. For example:

- Several PAs have very strong and localized maps to a selected set of ISO requirements.
- Some PAs have very sparse maps where the relationships between PA practices and ISO requirements are scattered over a large number of instances.
- One ISO 9001 requirement, 7.5.4 "Customer property," did not map to any CMMI practices.

A good example of a disaggregated matrix that shows localized mapping is the relationship of the CAR PA to ISO 9001 requirements as shown in Table 5.3. In matrices of this type we have eliminated all ISO requirements that do not have mappings to the selected CMMI PA(s). The column labeled "Total" shows the sums of all maps to each CMMI SP. The second row shows the number of SPs mapped by each ISO requirement. The complete ISO 9001 to CMMI map is given in Appendix D.

Table 5.2 Terminology Comparison

ISO 9000:2000	*CMMI*
Quality Management System (QMS), Quality Manual	Organization's Set of Standard Processes
Quality Plan	Project Plan, Software Development Plan, System Engineering Management Plan, Data Management Plan

Table 5.3 Mapping of CAR SPs to ISO 9001

		ISO 9001	*8.4*	*8.5.1*	*8.5.2*	*8.5.3*
CMMI		Total	2	5	5	5
CAR	SP 1.1	4	X	X	X	X
CAR	SP 1.2	4	X	X	X	X
CAR	SP 2.1	3		X	X	X
CAR	SP 2.2	3		X	X	X
CAR	SP 2.3	3		X	X	X

All CAR SPs map to ISO 9001 sections 8.4 "Analysis of data" and 8.5 "Improvement" subsections: "Continual improvement," "Corrective action," and "Preventive action."

What is this map telling us? The ISO 9001 standard requires its users to continually improve their processes and analyze measures quantitatively, understand the effectiveness and suitability of the QMS, and to identify improvement opportunities. CMMI, at CL/ML 5, expects an organization to have a quantitative understanding of the variation in the organizational processes and to continually improve them. Through an understanding of common causes of variation, defect prevention can be implemented. This is a major purpose of the CAR PA. The fact that CAR is also used for analyzing problems other than defects, it is not relevant in this case since the ISO requirement is centered on defect prevention. The ISO 9001 requirements for adjusting corrective and preventive actions to be appropriate to the effect of the problems were not explicitly mapped to CMMI, but those requirements can be easily accommodated. Thus, we can conclude that both frameworks are compatible in this area.

Similarly, all OT SPs map to ISO 9001 sections 6.2.1 "Human resources–General" and 6.2.2 "Human resources–Competence, awareness and training," as shown in Table 5.4.

Table 5.4 Mapping of OT SPs to ISO 9001

CMMI		ISO 9001	6.2.1	6.2.2
		Total	7	7
OT	SP 1.1	2	X	X
OT	SP 1.2	2	X	X
OT	SP 1.3	2	X	X
OT	SP 1.4	2	X	X
OT	SP 2.1	2	X	X
OT	SP 2.2	2	X	X
OT	SP 2.3	2	X	X

What is this map telling us? The ISO 9001 standard requires that staff competence be ensured through evaluating training needs, providing necessary training, evaluating training effectiveness, and keeping records. This is equivalent to the purpose of the OT PA, although the OT practices are more detailed and provide additional guidance for establishing and implementing training capabilities. Also, implementation of PP SP 2.5 Plan for Needed Knowledge and Skills (not shown in Table 5.4) on the organization's projects provides support for efficient and effective training. Based on the mapping in Table 5.4, we can conclude that both frameworks are compatible in this area.

Table 5.4 shows that the OT PA is satisfied by implementing only two ISO 9001 requirements, 6.2.1 and 6.2.2. Additional ISO 9001 section 6 requirements are satisfied by other CMMI practices, namely, IPM SP 1.3 and IPM SP 2.4 (not shown in Table 5.4). The ISO 9001 section 6.1 requirement is satisfied by implementing GP 2.3 in all PAs. This means that to satisfy the ISO 9001 section 6 requirements, a CMMI rated organization will have to implement the OT PA, the IPM PA, and GP 2.3 in all relevant PAs.

In contrast to the focused mappings between ISO 9001 and CMMI shown for the CAR and OT PAs, there are a number of ISO 9001 requirements where the mapping is distributed over widely scattered CMMI practices. Those ISO 9001 requirements include:

- 6.3 Infrastructure
- 6.4 Work environment
- 7.3.1 Design and development planning
- 7.4.1 Purchasing process

- 7.5.1 Control of production and service provision (somewhat more compact map)
- 7.5.3 Identification and traceability
- 7.5.5 Preservation of product
- 8.2.4 Monitoring and measurement of product

As we will see in chapter 6, this has an impact on process improvement efforts, because satisfaction of those ISO requirements will require implementation of many PAs, even though just a few SPs are mapped to the ISO requirements.

Let's now consider SPs that are mapped with 60 percent confidence. For example, Measurement and Analysis SP 1.3 "Specify how measurement data will be obtained and stored" and SP 1.4 "Specify how measurement data will be analyzed and reported" among other practices, map to ISO 9001 section 7.6 "Control of monitoring and measuring devices." As shown in Table 5.5, these practices map strongly to the requirement to determine the monitoring and measurement needed, but only map with 60 percent confidence to the requirement intended to ensure that measurements can be carried out with desired accuracy. By analyzing the complete maps, one can see that implementing the CMMI MA processes will satisfy most of these ISO requirements, but specific attention will have to be given in some areas to ensure that all measurements and monitoring requirements are fully satisfied.

Table 5.5 Example of Mapping with 60 Percent Confidence

ISO 9001 Section	*ISO 9001 "Shall"*	*CMMI PAs*	*CMMI Practices*	*Confidence*
7.6	Control of monitoring and measuring devices			
	Determine monitoring and devices needed	MA	SP 1.1	100
		MA	SP 1.2	100
		MA	SP 1.3	100
		MA	SP 1.4	100
		VAL	SP 1.3	100
		VER	SP 1.3	100
	Establish monitoring processes	MA	GP 2.2	60
		MA	GP 2.9	60
		MA	SP 1.3	60
		MA	SP 1.4	60
		VAL	GP 2.9	60
		VAL	SP 1.3	60
		VER	GP 2.9	60
		VER	SP 1.3	60

As indicated earlier, there are several instances in which ISO 9001 requirements are quite detailed and are mapped to many CMMI PAs and practices. For example, the ISO 9001 section 6.1 requirement for providing resources to implement and maintain the quality management system is satisfied by GP 2.3 in all CMMI PAs (100 percent confidence) and OPD SP 1.6 "Establish and maintain work environment standards" (60 percent confidence) where OPD SP 1.6 enables the organization and projects to benefit from common tools, training, and maintenance. ISO 9001 section 6.1 also requires the organization to provide the resources needed to enhance customer satisfaction. This requirement also maps to GP 2.3 but only with a confidence level of 30 percent because there are no explicit requirements in CMMI for customer satisfaction. Taken together, GP 2.3 in all PAs and OPD SP 1.6 will satisfy this ISO requirement provided that some additional consideration is given to customer satisfaction and work environment standards.

Now let's take a look at the PAs and SPs that did not map to ISO 9001. Two PAs that did not map at all are Decision Analysis and Resolution (DAR) and Risk Management (RSKM). This is not a surprise since there are no explicit requirements in ISO 9001 for formal decision analysis and risk management. Table 5.6

Table 5.6 SPs Not Mapped to ISO 9001

PA	*SP*	*Description*
CM	SP 1.2	Establish a Configuration Management System
IPM	SP 3.2, SP 3.3, SP 3.4	IPPD Addition
OPD	SP 1.4	Establish the Organization's Measurement Repository
OPD	SP 1.5	Establish the Organization's Process Asset Library
OPD	SP 2.1, SP 2.2, SP 2.3	IPPD Addition
OPP	SP 1.5	Establish Process-Performance Models
PI	SP 2.2	Manage Interfaces
PMC	SP 1.1	Monitor Project Planning Parameters
PMC	SP 1.2	Monitor Commitments
PMC	SP 1.3	Monitor Project Risks
PP	SP 2.2	Identify Project Risks
QPM	SP 1.2	Compose a Defined Process
QPM	SP 1.3	Select the Subprocesses That Will Be Statistically Managed
QPM	SP 2.4	Record Statistical Management Data
RD	SP 2.2	Allocate Product Component Requirements
SAM	SP 1.1	Determine Acquisition Type
SAM	SP 2.5	Transition Products

shows the SPs that did not map to any ISO requirement. Among the SPs that did not map are those associated with the IPPD addition since no specific aspect of integrated product development is required in ISO 9001. These gaps will have to be specifically addressed for organizations using their QMS as a basis for their organizational set of standard processes (OSSP).

Now let's consider mapping just the SPs to ISO 9001 requirements. A word of caution before we embark to that analysis: As we see in these tables, ISO 9001 requirements will map to portions of PAs, which means that, in general, multiple PAs may be needed to satisfy an ISO requirement. We will discuss this further in chapter 6.

We begin with the mapping of the Engineering PAs: RD, REQM, TS, PI, VAL, and VER. Take, for example, the RD and REQM PAs, shown in Table 5.7. Here we can see that RD and REQM SPs satisfy ISO 9001 requirements in sections 7.1 through 7.2.3, 7.3.2, and 7.5.3. (Other section 7.3, 7.4, and 7.5 requirements are satisfied by other practices.) RD SP 2.2 Allocate Product Component Requirements is not mapped and will have to be specifically addressed. Also, there are no RD or REQM hits in ISO 9001 sections 4, 5, 6, and 8.

Table 5.7 RD–REQM Mapping to ISO 9001

		ISO 9001	*7.1*	*7.2.1*	*7.2.2*	*7.2.3*	*7.3.2*	*7.5.3*
CMMI		Total	4	6	9	5	9	1
RD	SP 1.1	3		X		X	X	
RD	SP 1.2	5	X	X	X	X	X	
RD	SP 2.1	4	X	X	X		X	
RD	SP 2.2	0						
RD	SP 2.3	3	X	X			X	
RD	SP 3.1	3		X		X	X	
RD	SP 3 2	3	X	X			X	
RD	SP 3.3	2			X		X	
RD	SP 3.4	2			X		X	
RD	SP 3.5	3			X	X	X	
REQM	SP 1.1	2			X	X		
REQM	SP 1.2	1			X			
REQM	SP 1.3	1			X			
REQM	SP 1.4	1						X
REQM	SP 1.5	1			X			

The RD and REQM SPs map strongly to these ISO 9001 requirements:

- 7.1 Planning product realization
- 7.2.1 Determination of requirements related to product
- 7.2.2 Review of requirements related to product
- 7.2.3 Customer communication
- 7.3.2 Design and development inputs
- 7.5.3 Identification and traceability

What is this map telling us? Table 5.7 shows that the REQM PA contributes little to satisfying the ISO 9001 requirements. Except for the traceability required by ISO 9001 section 7.5.3 "Identification and traceability," the RD PA satisfies most other requirements in this area. Looking from ISO 9001 to CMMI RD SP 2.2 Allocate Product Component Requirements, we realize that the ISO standard does not distinguish between products and product components. As a result, specific steps in the allocation of requirements to product components will have to be addressed when using an ISO 9001-based QMS as the OSSP. On the other hand, CMMI is not very strong in specifically requiring the customer communication and feedback expected by ISO 9001 section 7.2.3. Note that a confidence level of only 30 percent is shown for that map. The CMMI-based process descriptions should address customer communication, including satisfaction and feedback to satisfy multiple ISO requirements in that area.

The TS and PI PA maps are shown in Table 5.8. The TS SPs map strongly to the ISO 9001 requirements in sections 7.3.3 "Design and development outputs" and 7.5.1 "Control of production and service provision." TS SPs map weakly to sections 7.4.1 "Purchasing process" and 7.4.2 "Purchasing information."

PI SPs map strongly to the following ISO 9001 requirements:

- 6.3 Infrastructure
- 6.4 Work environment
- 7.4.1 Purchasing process
- 7.5.1 Control of production and service provision
- 7.5.3 Identification and traceability
- 7.5.5 Preservation of product

As would be expected, mapping of the ISO 9001 postdelivery requirements to CMMI are only of moderate strength, requiring specific steps if a CMMI-rated organization wants to implement such requirements. It is interesting to note that PI SP 2.2 Manage Interfaces is not mapped to any ISO 9001 requirement. This doesn't mean that ISO is not concerned about interfaces but only that such a requirement is not specifically called for in that framework.

What is this map telling us? Table 5.8 shows us that the TS and PI PAs cover essentially different processes and that their interaction will depend on the life

Table 5.8 TS and PI Mapping to ISO 9001

		ISO 9001	6.3	6.4	7.3.3	7.4.1	7.4.2	7.5.1	7.5.3	7.5.5
CMMI		Total	1	1	8	2	1	9	1	1
TS	SP 1.1	2			X		X			
TS	SP 1.2	1			X					
TS	SP 2.1	1			X					
TS	SP 2.2	2			X			X		
TS	SP 2.3	1			X					
TS	SP 2.4	2			X	X				
TS	SP 3.1	2			X			X		
TS	SP 3.2	2			X			X		
PI	SP 1.2	2	X	X						
PI	SP 1.3	1						X		
PI	SP 2.1	1						X		
PI	SP 2.2	0								
PI	SP 3.1	2				X		X		
PI	SP 3.2	1						X		
PI	SP 3.3	2						X	X	
PI	SP 3.4	2						X		X

cycle used. ISO 9001 section 7.3.3 “Design and development outputs” maps to all TS SPs, but the map also indicates that CMMI provides a lot more guidance for product design and implementation. When dealing with purchased products (ISO 9001 sections 7.4.1 and 7.4.2), both TS and PI make some contributions. On the other hand, PI processes provides much stronger support to ISO 9001 section 7.5.1 “Control of production and service provision” than TS. Also, ISO is much more specific in requiring information about identification and preservation of products and product components as they are being developed and integrated. The ISO-based processes should address creation and management of interfaces required by PI SP 2.2 in their process descriptions to satisfy CMMI.

Now let’s consider the VAL and VER PAs. As shown in Table 5.9, they are mapped strongly to

- 7.3.5 Design and development verification
- 7.3.6 Design and development validation
- 7.5.2 Validation of processes for production and service provision
- 8.1 General (product conformity)
- 8.2.4 Monitoring and measurement of product

Table 5.9 VAL–VER Mapping to ISO 9001

	ISO 9001	6.3	6.4	7.3.1	7.3.3	7.3.5	7.3.6	7.4.3	7.5.2	7.5.3	7.6	8.1	8.2.1	8.2.4	8.3	8.4	8.5.2	8.5.3
CMMI	Total	2	2	2	2	8	5	1	5	2	2	5	1	13	1	2	1	1
VAL SP 1.1	4			X			X		X					X				
VAL SP 1.2	5	X	X				X		X					X				
VAL SP 1.3	5				X		X		X		X			X				
VAL SP 2.1	5						X		X				X	X		X		
VAL SP 2.2	4						X		X	X				X				
VER SP 1.1	4			X		X						X		X				
VER SP 1.2	5	X	X			X						X		X				
VER SP 1.3	5				X	X					X	X		X				
VER SP 2.1	2					X								X				
VER SP 2.2	2					X								X				
VER SP 2.3	3					X						X		X				
VER SP 3.1	4					X		X				X		X				
VER SP 3.2	7					X				X				X	X	X	X	X

What is this map telling us? These two PAs show significant parallelism and share several mappings to ISO 9001 requirements. The planning, preparation, and analysis of the obtained results for VAL and VER satisfy the same ISO 9001 requirements (such as infrastructure and work environment definition in ISO 9001 sections 6.3 and 6.4, respectively, and monitoring and measurement of product in ISO 9001 section 8.2.4). However, CMMI, with separate PAs dedicated to verification and validation, provides a lot more specific guidance than the ISO 9001 requirements in sections 7.3.5 "Verification" and 7.3.6 "Validation." CMMI-based processes, however, require attention in the area of dealing with the nonconforming products (ISO 9001 section 8.3) and, as mentioned earlier, customer satisfaction (ISO 9001 section 8.4). Also, ISO 9001 is much more specific in requiring the review of nonconformities identified in those processes (ISO 9001 section 8.5.2) and in taking appropriate actions (ISO 9001 section 8.5.3).

An example where ISO requirements impact many process areas is project management. Let's take a look at the Project Management PAs, as shown in Table 5.10 and Table 5.11. Here we see that, as expected, PP and PMC practices are supported by IPM practices and cover ISO requirements in sections 4, 6, 7, and 8. Most of the maps for these PAs address:

- 7.1 Planning product realization
- 7.3.1 Design and development planning

Table 5.10 Project Management PA Mapping to ISO 9001

	ISO 9001	*4.2.1*	*4.2.3*	*6.2.2*	*6.3*	*6.4*	*7.1*	*7.2.3*	*7.3.1*	*7.3.2*	*7.3.4*	*7.5.4*	*8.2.1*	*8.2.3*	*8.3*	*8.4*	*8.5.2*	*8.5.3*
CMMI	Total	1	2	2	2	2	6	1	21	1	6	2	1	3	3	2	3	3
PP SP 1.1	1								X									
PP SP 1.2	1								X									
PP SP 1.3	2						X		X									
PP SP 1.4	1								X									
PP SP 2.1	1								X									
PP SP 2.2	1								X									
PP SP 2.3	5	X	X				X		X			X						
PP SP 2.4	3				X	X			X									
PP SP 2.5	2			X					X									
PP SP 2.6	1								X									
PP SP 2.7	2						X		X									
PP SP 3.1	2						X		X									
PP SP 3.2	1								X									
PP SP 3.3	1								X									
PMC SP 1.1	0																	
PMC SP 1.2	0																	
PMC SP 1.3	0																	
PMC SP 1.4	2		X									X						
PMC SP 1.5	2												X			X		
PMC SP 1.6	1										X							
PMC SP 1.7	1										X							
PMC SP 2.1	5										X			X	X		X	X
PMC SP 2.2	4													X	X		X	X
PMC SP 2.3	4													X	X		X	X
IPM SP 1.1	2						X		X									
IPM SP 1.2	1									X								
IPM SP 1.3	2				X	X												
IPM SP 1.4	2						X		X									
IPM SP 1.5	1								X									
IPM SP 1.6	1															X		
IPM SP 2.1	3							X	X		X							
IPM SP 2.2	2								X		X							
IPM SP 2.3	2								X		X							
IPM SP 3.1	1			X														
IPM SP 3.2	0																	
IPM SP 3.3	0																	
IPM SP 3.4	0																	
IPM SP 3.5	1								X									

- 7.3.4 Design and development review
- 8.2.3 Monitoring and measurement of process (take corrective action)
- 8.3 Control of nonconforming product (take action to preclude use)
- 8.4 Analysis of data
- 8.5.2 Corrective action
- 8.5.3 Preventive action

What is this map telling us? The PP PA fully satisfies the requirements in ISO 9001 section 7.3.1 "Design and development planning." ISO 9001 section 7.1 "Planning product realization" provides requirements for the project plan that is called for in PP SP 2.7. However, the PP PA provides more helpful guidance for developing and documenting the plan. The PMC PA maps to those ISO 9001 requirements that deal with reviews, such as section 7.3.4, and taking corrective actions, such as section 8.2.3. It is important to note that PMC SP 1.1, SP 1.2, and SP 1.3 are not mapped at all. Those SPs deal with monitoring project planning parameters, commitments, and risks, respectively. This implies that an ISO 9001-registered organization will have to put a special emphasis on the monitoring processes. On the other hand, as noted earlier, a CMMI-rated organization will have to emphasize monitoring of and reacting to customer satisfaction status. The IPM PA contributes very little beyond the mapping to ISO requirements provided by PP and PMC except for the mapping to ISO section 7.3.2 that requires organizations to consider previous similar designs.

As shown in Table 5.11, two SAM SPs were not mapped at all: SP 1.1 Determine Acquisition Type and SP 2.5 Transition Products. These practices will have to be specifically addressed for ISO-registered organizations that are striving to achieve

Table 5.11 SAM Mapping to ISO 9001

		ISO 9001	*4.1*	*7.4.1*	*7.4.2*	*7.4.3*	*8.2.4*	*8.4*
CMMI		Total	4	6	1	5	1	2
SAM	SP 1.1	0						
SAM	SP 1.2	1		X				
SAM	SP 1.3	4	X	X	X	X		
SAM	SP 2.1	3	X	X		X		
SAM	SP 2.2	4	X	X		X		X
SAM	SP 2.3	4	X	X		X		X
SAM	SP 2.4	3		X		X	X	
SAM	SP 2.5	0						

CMMI capability or maturity levels. However, there is a nearly complete mapping to 7.4.1 "Purchasing process" and 7.4.3 "Verification of purchased product."

Let's now revisit the support processes. At the beginning of this section we mentioned that CAR, one of the Support PAs, satisfied ISO 9001 sections 8.4 and 8.5. We look next at the Measurement and Analysis (MA) PA shown in Table 5.12. An organization rated at CMMI ML 2 or above will have this PA implemented and would therefore satisfy most of the requirements in ISO 9001 section 8.

What is this map telling us? One would expect that the MA process area and ISO 9001 section 8, Measurement, analysis and improvement, would align well. However, there are four requirements in this section of the ISO standard that deal with improvement that are not satisfied by MA implementation but are instead satisfied by OPF (ISO 9001 sections 8.2.2, 8.5.2, and 8.5.3), PMC (ISO 9001 sections 8.3 and 8.5.3), and PPQA (ISO 9001 sections 8.5.2 and 8.5.3). Those PAs are not shown in this table. This implies that to fully satisfy all the requirements of ISO 9001 section 8, in addition to MA, these three PAs will have to be implemented, at least to the CL1 level. Here, too, CMMI provides extensive guidance for performing measurement and analysis that an ISO 9001-registration bound organization may be able to leverage.

Let's now take a look at the remaining two Support PAs: CM and PPQA, shown in Table 5.13. The CM PA maps to ISO 9001 sections 7.3.7 "Control of design and development changes," 7.5.1 "Control of production and service provision," and 7.5.3 "Identification and traceability." In addition, CM maps to section 4.2.3 "Control of documents," and PPQA maps to section 4.2.1 "General" (records) dealing with control of documentation. PPQA maps to sections 8.8.2 "Internal audit," 8.2.4 "Monitoring and measurement of product," 8.3 "Control of nonconforming

Table 5.12 MA Mapping to ISO 9001

		ISO 9001	*5.4.1*	*7.2.2*	*7.5.3*	*7.6*	*8.1*	*8.2.1*	*8.2.3*	*8.2.4*	*8.4*	*8.5.1*
CMMI		Total	1	1	1	4	3	3	4	2	8	2
MA	SP 1.1	6	X	X		X		X			X	X
MA	SP 1.2	5				X	X	X	X		X	
MA	SP 1.3	4				X	X		X		X	
MA	SP 1.4	4				X	X		X		X	
MA	SP 2.1	2								X	X	
MA	SP 2.2	5						X	X	X	X	X
MA	SP 2.3	1									X	
MA	SP 2.4	2			X						X	

Table 5.13 CM and PPQA Mapping to ISO 9001

		ISO 9001	*4.2.1*	*4.2.3*	*4.2.4*	*7.3.7*	*7.5.1*	*7.5.3*	*8.1*	*8.2.2*	*8.2.4*	*8.3*	*8.5.2*	*8.5.3*
CMMI		Total	1	3	1	3	1	4	1	4	1	5	2	1
CM	SP 1.1	2		X				X						
CM	SP 1.2	0												
CM	SP 1.3	2					X					X		
CM	SP 2.1	3				X		X				X		
CM	SP 2.2	4		X		X		X				X		
CM	SP 3.1	3				X		X				X		
CM	SP 3.2	2		X							X			
PPQA	SP 1.1	1								X				
PPQA	SP 1.2	2							X	X				
PPQA	SP 2.1	4								X		X	X	X
PPQA	SP 2.2	3	X		X					X			X	

product," and the closely related 8.5.2 "Corrective action" and 8.5.3 "Preventive action."

What is this map telling us? These two PAs overlap only through ISO 9001 section 8.3 "Control of non-conforming product," where PPQA identifies nonconformities and CM keeps track of them. It is important to note that CM SP 1.2 Establish a Configuration Management System is not mapped which means that an ISO 9001-registered organization that is pursuing a CMMI rating will have to document how it will establish and maintain a system for controlling work products. It is not surprising that by implementing ISO 9001 section 8.2.2 "Internal audit," the PPQA PA will be fully satisfied. However, CMMI is a process model that addresses processes that when implemented will lead toward quality products and thus explicitly requires both process and product reviews.

Let's turn our attention to the organizational-level PAs. The OT PA was discussed earlier in this section. The OPD and OPF maps are shown in Table 5.14.

What is this map telling us? In CMMI, the OPD and OPF PAs interact quite heavily. However, several sections in ISO 9001 address those CMMI processes but with little or no interaction. These two PAs support the process improvement infrastructure that is required by ISO 9001 section 4.1 (establish quality management system, identify processes, determine sequences, etc.). Specifically, OPD is needed for establishing a documented quality policy and quality manual (section 4.2.1, document quality manual and procedures), whereas the OPF PA supports ISO 9001 requirements for quality policy (section 5.3), quality objectives (section 5.4.1), and quality management system planning (section 5.4.2). It is important

Table 5.14 Organizational PAs Mapping to ISO 9001

CMMI		*ISO 9001*	*4.1*	*4.2.1*	*4.2.2*	*5.3*	*5.4.1*	*5.6.1*	*6.1*	*6.3*	*6.4*	*7.1*	*8.1*	*8.2.2*	*8.4*	*8.5.1*	*8.5.2*	*8.5.3*
		Total	3	3	1	1	1	2	1	1	1	3	4	6	2	9	3	1
OPD	SP 1.1	4	X	X	X							X						
OPD	SP 1.2	2		X								X						
OPD	SP 1.3	2		X								X						
OPD	SP 1.4	0																
OPD	SP 1.5	0																
OPD	SP 1.6	3							X	X	X							
OPD	SP 2.1	0																
OPD	SP 2.2	0																
OPD	SP 2.3	0																
OPF	SP 1.1	4				X	X							X		X		
OPF	SP 1.2	5						X					X	X	X	X		
OPF	SP 1.3	5						X					X	X	X	X		
OPF	SP 2.1	4											X	X		X	X	
OPF	SP 2.2	5	X										X	X		X	X	
OPF	SP 3.1	2														X	X	
OPF	SP 3.2	1														X		
OPF	SP 3.3	3	X											X		X		
OPF	SP 3.4	2														X		X

to note that OPD SP 1.4 Establish the Organization's Measurement Repository and SP 1.5 Establish the Organization's Process Asset Library do not have ISO 9001 equivalents. An ISO 9001-registered organization that would like to achieve CMMI CL/ML 3 or above would need to establish those two repositories. The OPD specific goal SG 2 Enable IPPD Management is not addressed since ISO 9001 does not specifically deal with the IPPD management method.

The analysis of high maturity PAs (CAR, OID, OPP, and QPM) and their maps to ISO 9001 can become somewhat involved. These PAs map mostly to section 8 "Measurement, analysis and improvement," as shown in Table 5.15.

ISO 9001 has no explicit requirements for statistical process control; organizations will need to determine which analytical methods will be used to analyze obtained data. The selected methods may include statistical methods. Therefore, both CAR and OID, as shown in the maps, do not presuppose the stable, quantitatively managed processes expected by CMMI—they address problems and their resolution. The process performance baseline that is mapped to ISO 9001 section 8.4 is used as a tool for analyzing process and products trends rather

Table 5.15 Mapping of High Maturity PAs to ISO 9001

		ISO 9001	*4.1*	*5.4.1*	*7.1*	*8.1*	*8.2.3*	*8.4*	*8.5.1*	*8.5.2*	*8.5.3*
CMMI		Total	1	2	1	9	2	11	12	5	5
OPP	SP 1.1	1						X			
OPP	SP 1.2	2					X	X			
OPP	SP 1.3	2		X				X			
OPP	SP 1.4	1						X			
OPP	SP 1.5	0									
QPM	SP 1.1	2		X	X						
QPM	SP 1.2	0									
QPM	SP 1.3	0									
QPM	SP 1.4	1						X			
QPM	SP 2.1	1				X					
QPM	SP 2.2	1				X					
QPM	SP 2.3	1					X				
QPM	SP 2.4	0									
CAR	SP 1.1	4						X	X	X	X
CAR	SP 1.2	4						X	X	X	X
CAR	SP 2.1	3							X	X	X
CAR	SP 2.2	3							X	X	X
CAR	SP 2.3	3							X	X	X
OID	SP 1.1	3				X		X	X		
OID	SP 1.2	3				X		X	X		
OID	SP 1.3	3				X		X	X		
OID	SP 1.4	3				X		X	X		
OID	SP 2.1	3	X			X			X		
OID	SP 2.2	2				X			X		
OID	SP 2 3	2				X			X		

than for analyzing an organization's data to establish a distribution and range of results that characterize the expected performance of the selected subprocesses. This consideration is important for those CMMI-rated organizations that have not implemented the high maturity PAs or generic goals GG 4 and GG 5. Those organizations will be able to achieve ISO 9001 registration provided that they develop additional processes and procedures to satisfy the requirements of ISO 9001 section 8.

What is this map telling us? The CAR mapping was discussed earlier in this chapter. All OID SPs are mapped to ISO 9001 requirements. The mapping shows that the ISO 9001 requirement for continual process improvement and the purpose of OID coincide. However, the quantitative understanding of process performance in support of quality and quantitative management of projects that is expected by CMMI is absent in ISO 9001 (QPM SP 1.2 and SP 1.3 are not mapped). ISO 9001 requirements for establishing quality objectives (sections 5.4.1 and 7.1) are satisfied by QPM SP 1.1 and OPP SP 1.3, although in CMMI, those objectives are based on a quantitative knowledge of process performance and product quality. Also absent from ISO 9001 are CMMI practices for establishing process performance models (OPP SP 1.5), composing a project's defined process based on historical information (QPM SP 1.2), and selecting subprocesses that will be statistically managed (QPM SP 1.3). An ISO 9001-registered organization will have to revisit its approach to quantitative project management and espouse some of the guidelines given in CMMI.

Thus far, the discussions in this chapter have not addressed CMMI GPs. The GPs play an important role in process improvement, specifically in institutionalizing processes implementation. When SPs and GPs are mapped to an ISO 9001 framework, they will, in general, map to several requirements. In ISO 9001 those requirements are unrelated and may belong to different sections and represent different topics.

Let's now consider an example of mapping some SPs and GPs to the ISO framework as shown in Figure 5.4. Here we see that CMMI SP *x* in PA *M* is related to GP *y* in PA *N*. This is similar to the relationships between PAs and GPs given in Table 7.2 of the CMMI book (Chrissis 2007), which we discussed in chapter 3. When we look at the CMMI–ISO 9001 mapping, SP *x* is shown to map to ISO

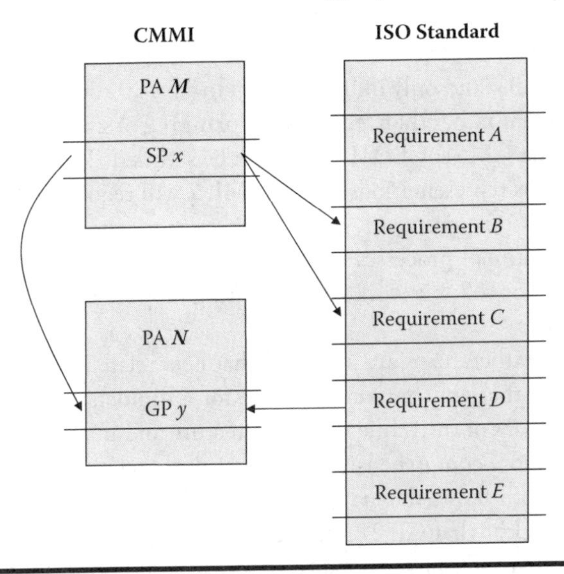

Figure 5.4 Conceptual SP/GP mapping.

requirements *B* and *C*, and GP *y* is mapped to ISO requirement *D*. This situation is not uncommon; where there is a close relationship between PAs and GPs in CMMI, there may be no similar relationships in the ISO standard. The users of these maps will have to understand both frameworks and take their inner relationships into account. Let's take a few examples.

GP 2.5 Train People maps to:

- ISO 9001 section 6.2.1 "General"—This section also maps to all OT practices, as shown in Table 5.4.
- ISO 9001 section 6.2.2 "Competence, awareness and training" —This section also maps to PP SP 2.5.

This is a "consistent" map because it indicates the relationship between GPs and SPs in CMMI through ISO requirements.

Now let's take GP 2.7 Identify and Involve Relevant Stakeholders, which maps to:

- ISO 9001 section 7.2.3 "Customer feedback," which also maps to IPM SP 2.1 Manage Stakeholder Involvement, as shown in Table 5.10
- ISO 9001 section 7.3.1 "Manage interfaces," which also maps to all SPs in IPM SG 2 Coordinate and Collaborate with Relevant Stakeholders
- ISO 9001 section 7.3.4 "Appropriate functions participate in reviews," which also maps to IPM SG 2 Coordinate and Collaborate with Relevant Stakeholders

It is surprising that PP SP 2.6 and PMC SP 1.5 do not map to ISO 9001 requirements 7.2.3 and 7.3.4 and only indirectly map to 7.3.1 (through the update of plans during development rather than through the planning of stakeholder involvement), which implies that CMMI CL/ML 2 may not be satisfied. Those organizations that use the continuous representations or are at ML 2 will require additional processes to satisfy these ISO requirements. In addition, even for those organizations at CL/ML 3, some additional processes may be required because the CMMI mapping to ISO 9001 section 7.2.3 is weak, concentrating on only one kind of stakeholder, namely, customers.

As indicated earlier, maps are just tools that help relate the requirements of one framework to another and cannot substitute for an understanding of each framework and the impact of the framework architecture and intent.

To obtain a more comprehensive picture of those relationships let's take a look at Table 5.16, which shows the mapping of the GPs in a typical PA. By inspection, it becomes quite clear that GP 2.1 through GP 3.2 map to a large number of ISO requirements. That, of course, is expected because although ISO 9001 does not explicitly require institutionalization, it is strongly implied.

Table 5.16 Example Mapping of GPs in a Typical PA

CMMI	*ISO 9001*	*4.1*	*4.2.1*	*4.2.2*	*4.2.3*	*4.2.4*	*5.1*	*5.3*	*5.5.1*	*5.6.2*	*5.6.3*	*6.1*	*6.2.1*	*6.2.2*	*6.3*	*6.4*	*7.1*	*7.2.3*	*8.1*	*8.2.3*	*8.2.4*	*8.4*	*8.5.1*
	Total	4	1	1	1	1	3	1	1	1	2	1	1	1	1	1	3	1	1	1	1	1	2
PP GP 2.1	5	X	X				X	X															X
PP GP 2.2	2			X													X						
PP GP 2.3	6	X					X				X	X			X	X							
PP GP 2.4	1								X														
PP GP 2.5	2												X	X									
PP GP 2.6	3				X	X											X						
PP GP 2.7	1																	X					
PP GP 2.8	4	X																	X	X	X		
PP GP 2.9	1	X																					
PP GP 2.10	4						X			X	X												X
PP GP 3.1	1																X						
PP GP 3.2	1																					X	

Nevertheless, an ISO 9001-registered organization would have to undertake significant steps to achieve a solid CMMI CL/ML 3 rating. In ISO 9001 there is no concept of "tailoring," which is a major enabler of GP 3.1. IPM SP 1.1 and the OPD SPs that relate to GP 3.1 map to:

- 4.1 General requirements (identify processes, determine sequence)
- 4.2.1 General (documenting Quality Manual)
- 4.2.2 Quality manual (scope process interactions)
- 7.1 Planning of product realization
- 7.3.1 Design and development planning (determine life cycle stages, update plans)

GP 3.1, however, is only mapped to ISO 9001 section 7.1 and is related to planning rather than to establishing a project's defined process. An ISO 9001-registered organization may have to reevaluate its process approach by expanding the project planning process to include development and implementation of some standard processes that will be used on projects.

In the discussion so far, we have not considered the strength of each mapping, expressed as certainty or confidence. Let's now consider the GPs that mapped to ISO 9001 requirements at a 30 percent confidence level, recalling that 30 percent in mapping confidence means that the ISO requirement does not correspond completely to some CMMI practice, but with effort it may be interpreted and implemented in CMMI. For example, GP 2.7 "Identify and involve the relevant stakeholders of the process as planned" maps to ISO 9001 section 7.2.3 "Customer communication" and specifically to the implementation of customer feedback. From our previous discussion we realize that there is very little in CMMI that deals explicitly with customer communication except through the involvement of stakeholders. Through awareness of this ISO requirement, an organization that has or is developing its stakeholder management process can require projects to communicate with their customers (the major stakeholders in product development) and collect their input, paying special attention to customer complaints. In a similar fashion, some other specific customer communications may be implemented through processes related to IPM SP 2.1 "Manage the involvement of the relevant stakeholders in the project." Other ISO requirements related to section 7.2.3 are fully satisfied with implementation of the RD and REQM processes.

Going in the opposite direction, an ISO 9001-registered organization will have to add a stakeholder management process to satisfy CMMI requirements. In some cases, ISO 9001 requirements are much more precise than CMMI, thereby requiring that CMMI-based processes specifically address those requirements. For example, one of the requirements in ISO 9001 section 8.2.4 deals with the fact that a product should not be released until planned activities are complete. This requirement maps to CM SP 3.2 "Perform configuration audits to maintain integrity of the configuration baselines" with 30 percent confidence requiring the corresponding CMMI-based CM audit process to specifically address this requirement. In addition to this example, there are an additional 13 instances of mapping with 30 percent confidence dealing with monitoring customer perception, analyzing customer perception data, and verifying that corrective actions were taken. There are also several requirements dealing with the ability to meet product requirements and dealing with quality records that would require similar actions on the part of an organization wishing to satisfy those ISO 9001 requirements when implementing CMMI-based processes.

When mapping confidence rises to 60 percent, the interpretation of the frameworks is more subtle, requiring fewer additional details when developing and implementing CMMI-based processes to satisfy ISO 9001.

For example, GP 2.1 "Establish and maintain an organizational policy for planning and performing the <process name> process" maps to ISO 9001 section 4.1 "General requirements" and section 5.1 "Management commitment." Those requirements require an organization to manage its processes according to ISO 9001, ensure that quality objectives are established, and communicate the importance of meeting customer and other requirements to the staff. An organization that already has or is developing CMMI-based policies will find it easy to extend those policies to specifically address these ISO 9001 requirements and ensure that when the requirements are implemented, the staff is informed of their importance.

ISO 20000:2005 to CMMI Maps

On the surface, when one looks at the ISO 20000-to-CMMI map (and its inverse), it may appear as if those two frameworks are not very compatible. This is no surprise, because they cover different domains: CMMI addresses development processes, whereas ISO 20000 addresses service delivery and management. So why did we decide to include ISO 20000 in this book? There are several reasons. First, the popularity of both of these frameworks and their extensive use in information technology (IT) organizations was an important reason for considering their comparison. The second reason is to take advantage of the use of multiple frameworks for process improvement. Can implementation of one framework help an organization successfully implement another based on their synergy? The third reason addresses the issue of compliance. Can implementation of one framework satisfy the requirements of both frameworks? If not, how much additional effort would be required? We will address the second issue in chapter 6 and the third issue in chapter 7. The SEI, government, and industry are developing the CMMI for Services (CMMI-SVC), which may help resolve many issues discussed in this section.

As with the other examples of framework maps described in this chapter, the purpose of the ISO 20000-to-CMMI map is to determine the synergy between the ISO standard and CMMI and then, based on this synergy, determine where additional CMMI interpretation is needed.

The synergy between these two frameworks and their specific strengths is shown in Figure 5.5. Because many organizations use both ISO 9001 and ISO 20000, with or without CMMI, we will also address the compatibility of those two ISO frameworks in chapter 6.

Like the whole family of ISO standards, ISO 20000:2005, Part 1 is based on the integrated process approach described in chapter 4 and specifies requirements

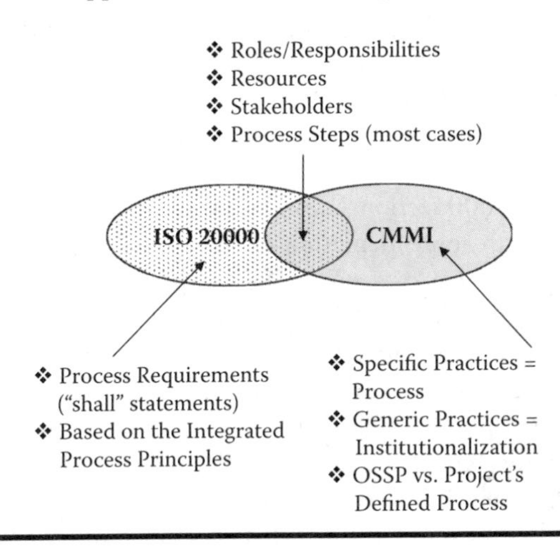

Figure 5.5 ISO 20000 and CMMI synergy.

in the form of "shall" statements. CMMI, at maturity or capability level 3, advocates development of an OSSP that can be tailored to the individual project's need. The OSSP provides an integrated process with artifacts, or assets, which can be found in a process asset library and reused by projects. CMMI SPs describe the characteristics of an idealized process that is institutionalized by the implementation of GPs. What both ISO 20000 and CMMI have in common is the need for defining process resources, roles, and responsibilities; proactively identifying and involving the stakeholders; and defining process steps rather than describing how to implement those steps.

We will see later in this chapter that an OSSP that can satisfy ISO 20000 requirements will incorporate many CMMI PAs with no more than minor modifications or interpretations, although some ISO requirements may require development of additional processes. For example, in an organization that is CMMI ML 3 rated:

- The process infrastructure will already exist (based on the OPF, OPD, OT PAs that will, in general, satisfy ISO 20000 section 3.2 "Documentation requirements" and 4.4 "Continual improvement").
- The processes associated with the Project Management PAs (PP, PMC, SAM, RSKM, and IPM) and Support PAs (PPQA, MA, and CM) will be carried over with little or no modification.
- The Engineering PAs (TS, PI, and, to some extent, VER and VAL) will have to be replaced. RD and REQM can be implemented with the understanding that the notion of Service Level Agreements (SLA) and Service Level Management (SLM) will have to be accommodated.

If we translate this to the effort required to update the OSSP, it means that the processes needed to satisfy Service Level Management (ISO 20000 section 6), Business Relations and Supplier Management (ISO 20000 section 7), Incident and Problem Management (ISO 20000 section 8), and Release Management (ISO 20000 section 10) will have to be written and their impact on other processes will have to be assessed so that the appropriate interfaces among those processes may be established. (ISO 20000 section 9 Configuration and Change Management will be mostly satisfied with implementation of the CM PA.) By capitalizing on the PA–GP interactions shown earlier, most GPs can be satisfied. The appraisal impacts of such "process reengineering" will be discussed in chapter 7.

So, where do those two frameworks diverge? As we have seen earlier in this chapter, one difference between ISO standards and CMMI is in the depth of coverage in process descriptions. Also, their applicable domains differ: ISO 20000 addresses service management whereas CMMI addresses development. Table 5.17 summarizes the types of differences found between the two frameworks. In the following sections we will describe those differences and will outline the conclusions that can be drawn from the maps.

Table 5.17 Summary of ISO 20000 and CMMI Differences

ISO 20000	*CMMI*
ISO section matches CMMI PA well (example: 9.1 CM)	CMMI PA matches ISO section well (example: CM)
No match found in CMMI (example: 8.2 Incident management)	No match found in ISO 20000 (example: DAR)
Knowledge area addressed in single section (example: 4.1 Plan service management)	Area addressed in multiple PAs—more detail in CMMI (mapping: CM, PP, PMC, PPQA, RD, REQM, RSKM)
No explicit notion of institutionalization	Generic Goals and Generic Practices

As we have done for all of the other maps, a detailed map between ISO 20000 and CMMI was developed using ISO 20000 as the primary framework. For each shall statement in the ISO standard, the appropriate CMMI SPs or GPs were identified that, when implemented, would satisfy that ISO requirement. The inverse map, from CMMI to ISO 20000, was automatically generated by the tool.

The shall statements in ISO 20000 have many substatements (for example, enumerated lists that further delineate specific requirements). For clarity, we have generally folded the mappings to those substatements into the higher-level statements. However, we kept track of the number of matches between the ISO standard and CMMI in the detailed maps stored in the mapping tool.

Because different domains are covered by ISO 20000 and CMMI, we have relied on the confidence factors explained earlier in this chapter when drawing conclusions on the synergy between the frameworks.

Let's take a look at some examples of the use of confidence factors. ISO 20000 states that, when implementing a new or changed service, the expected outcome from operating this new or changed service will have to be addressed and determined in measurable terms. There is nothing in CMMI that directly addresses this particular requirement, since it is primarily a postdelivery activity, which is not addressed in CMMI. The requirement can, however, be interpreted by using QPM SP 1.1 Establish the Project's Objectives, subpractice 4 "Define and document measurable quality and process-performance objectives for the project." We considered this to be a weak map and, therefore, selected a confidence factor of 30, meaning that considerable effort would be required when developing a CMMI-based process to satisfy this requirement.

Another example is the widespread ISO 20000 requirement for specifically addressing service management. For those cases, we usually selected a confidence factor of 60 because, although CMMI does not specifically address service management, the service management process can be planned and implemented using CMMI PAs with rather small interpretations or modifications.

An example of an area where we selected a confidence factor of 0, or No Map, is service continuity since CMMI does not directly deal with processes occurring after product delivery; interpreting CMMI practices to meet such a requirement would be a stretch. Of course, appropriate processes could be developed and incorporated in the organization's set of standard processes. Indicating that no map exists raises a flag to process developers to turn their attention to such special processes.

In contrast, as an example of the Full Map, consider the ISO 20000 requirement that management shall "manage risks to the service management organization and services," which maps quite well to GP 2.10 in the Risk Management PA.

As mentioned earlier, maps are very subjective and different people will create similar but different maps. For example, consider the ISO 20000 requirement for creating a capacity plan. This requirement could be mapped to PP SP 2.4 Plan for Project Resources, because the project plan should address the project resource requirements. Others may see this requirement as mapping to GP 2.3 in all relevant PAs. In this case, we have chosen to map to PP SP 2.4, which is associated with GP 2.3, as indicated by Chrissis (2006).

As an example of mapping, a small portion of the map related to the RD PA is shown in Table 5.18 showing how ISO 20000 sections 3.1, 3.2, and 3.3 map to the RD PA (note that there are also maps to additional PAs that are not shown). This example shows the requirements in the section identified as R1, R2, R3, and so forth, and elements of an enumerated list are shown as a, b, c, etc. In the "Total" row, we indicate the number of matches in the columns below so that we can see at a glance when and how frequently an ISO requirement relates to CMMI practices. This example shows that, as is the case in for most of the maps, these are not one-to-one maps. An example of the rolled-up matrix for the RD PA is shown in Table 5.19.

We will now discuss the maps in a fashion similar to that shown earlier with the maps between ISO 9001 and CMMI. By examining the CMMI-to-ISO 20000 maps, we see that DAR is the only PA that did not map at all. A further analysis of the mapping matrix points to several trends. Since CMMI-DEV does not address postdelivery processes, there are specific ISO 20000-required processes that will not have CMMI equivalents:

- 7.2 Business relationship management—Complaint process; customer satisfaction and business relationship processes
- 7.3 Supplier management—Contractual disputes process
- 8.2 Incident management—All aspects of incident management
- 8.3 Problem management—Problem resolution including prevention; relationship of problem management to incident management
- 10.1 Release management—All aspects of the release management process including emergency releases, reversing release if not successful, measuring success or failure, and improving the service

Table 5.18 CMMI-to-ISO 20000 Mapping Matrix Example

	ISO 20000	3.1								3.2						3.3		
		R1	*R2a*	*R2b*	*R2c*	*R2d*	*R2e*	*R2f*	*R2g*	*R1*	*R2a*	*R2b*	*R2c*	*R2d*	*R3*	*R1*	*R2*	*R3*
CMMI	Total	0	0	0	1	0	0	0	0	0	2	2	0	0	0	1	1	0
RD SP 1.1	0																	
RD SP 1.2	1										X							
RD SP 2.1	1										X							
RD SP 2.2	0																	
RD SP 2.3	1																	
RD SP 3.1	0																	
RD SP 3.2	0																	
RD SP 3.3	0																	
RD SP 3.4	0																	
RD SP 3.5	0																	
RD GP 1.1	0																	
RD GP 2.1	1											X						
RD GP 2.2	1											X						
RD GP 2.3	1																	
RD GP 2.4	2															X		
RD GP 2.5	1																X	
RD GP 2.6	0																	
RD GP 2.7	0																	
RD GP 2.8	0																	
RD GP 2.9	0																	
RD GP 2.10	1				X													
RD GP 3.1	1																	
RD GP 3.2	0																	

In addition there are some very weak maps such as ISO 20000 section 3.1 "Management responsibility." This section does not map to any CMMI SPs, but maps weakly to GP 2.3, GP 2.4, and GP 2.10. It also requires that GP 2.1 be written to explicitly address a policy for service management.

On the other hand, as shown in the examples in Table 5.20, there are several very strong relationships between ISO 20000 and CMMI processes. Going in the other direction (from CMMI to ISO 20000), there are some interesting mapping results. For example, there are some PAs where very few SPs map to ISO 20000,

Table 5.19 CMMI-to-ISO 20000 Mapping Matrix, Rolled-Up

	ISO 20000	*3.1*	*3.2*	*3.3*	*4.1*	*4.2*	*4.3*	*4.4.1*	*4.4.2*	*4.4.3*	*5*	*6.1*	*6.2*	*6.3*	*6.4*	*6.5*	*6.6*	*7.2*	*7.3*	*8.2*	*8.3*	*9.1*	*9.2*	*10.1*
CMMI	Total	1	4	2	4	5	5	1	0	0	5	12	0	3	0	0	0	2	1	0	0	0	0	0
RD SP 1.1	3											2						1						
RD SP 1.2	4		1									1		1				1						
RD SP 2.1	4		1								1	1		1										
RD SP 2.2	1													1										
RD SP 2.3	2				1														1					
RD SP 3.1	0																							
RD SP 3.2	0																							
RD SP 3.3	2										1	1												
RD SP 3.4	2										1	1												
RD SP 3.5	2											2												
RD GP 2.1	2		1			1																		
RD GP 2.2	2		1			1																		
RD GP 2.3	3				1	1					1													
RD GP 2.4	4			1	1	1					1													
RD GP 2.5	1			1							0													
RD GP 2.6	1										0	1												
RD GP 2.7	2										0	2												
RD GP 2.8	4						3				0	1												
RD GP 2.9	1							1			0													
RD GP 2.10	3	1					2				0													
RD GP 3.1	2				1	1					0													
RD GP 3.2	0										0													

as shown in Table 5.21. Table 5.22 shows that there are some PAs where the SPs map strongly to a small number of ISO requirements. Finally, there are several ISO 20000 sections that map to a wide range of PAs, such as CM, MA, PMC, and RSKM.

Let's start our detailed analysis with the Causal Analysis and Resolution (CAR) PA. The complete ISO 20000 to CMMI map is given in Appendix G. The CAR portion of the map is shown in Table 5.23. As mentioned earlier in this chapter, the ISO 20000–CMMI relationship matrices show rolled up results for clearer presentation

Table 5.20 Strong ISO 20000–CMMI Relationships

	ISO 20000 Requirement	*Maps to CMMI*
3.3	Competence, awareness and training	OT, PMC, PP Except: The need for the staff awareness of their actions and contributions to service management objectives
4.1	Plan service management (Plan)	IPM, PP, RD, RSKM Except: Need to explicitly cover service management, quality of service management, and other service-specific issues
4.2	Implement service management and provide the service (Do)	IPM+IPPD Addition, PMC, PP, RSKM Except: Need to explicitly cover service management, managing teams (e.g., service desk)
4.3	Monitoring, measuring and reviewing (Check)	MA, OPF, PMC, PPQA Except: Need to explicitly cover service management
4.4.2	Management of improvements	OID, OPD, OPF Except: Need to explicitly address services
4.4.3	Activities	MA, OID, OPD, OPF, OPP Except: need to explicitly cover service management
9.1	Configuration management	CM Except: Need to integrate configuration and change management, interfacing financial asset management, service management effectiveness monitoring
9.2	Change management	CM, OPF Except: Need to define scope of service and infrastructure changes, classification of requests, reversing unsuccessful changes, control of emergency changes, reviewing changes to detect trends

and easier understanding. In this case, ISO 20000 section 8.3 has eight individual shall-level requirements, six of which map to four CAR SPs.

What is this map telling us? ISO 20000 section 8.3 "Problem management" maps to most CAR SPs indicating synergy between CMMI and ISO. However,

Table 5.21 CMMI PAs with Few Maps to ISO 20000

CMMI PA	Specific Practices	Map to ISO 20000	
PI	SP 2.1, SP 2.2	7.3	Supplier management
TS	SP 2.3	7.3	Supplier management
OPP	SP 1.3	4.4.3	Activities
		7.2	Business relationship management
QPM	SP 1.1	5	Planning or implementing new or changed service
		7.2	Business relationship management
	SP 2.2, SP 2.3	6.5	Capacity management

Table 5.22 CMMI PAs with Focused Mapping to Few ISO 20000 Requirements

CMMI PA	Specific Practices	Map to ISO 20000	
CAR	All SPs except SP 1.1	8.3	Problem management
OID	All SPs except SP 1.3	4.4.2	Management of improvements
	All SPs	4.4.3	Activities
OPD	SP 1.1	3.2	Documentation requirements
		4.4.2	Management of improvements
		4.4.3	Activities
		7.3	Supplier management
		10.1	Release management
OT	SP 1.1, SP 2.1	3.3	Competence, awareness, and training
	SP 1.1, SP 1.2, SP 1.3	5	Planning or implementing new or changed service
RSKM	All SPs	4.2	Implement service management and provide the services (Do)
		6.6	Information security management
	SP 1.1, SP 1.2, SP 1.3	4.1	Plan service management (Plan)

there are additional PAs that are mapped to this section, namely, CM SP 2.1, SP 2.2, MA SP 2.1, and OID SP 2.1 that are not shown in Table 5.23. Since handling of problems encountered after delivery is not covered in CMMI, that map is somewhat weaker. Hence, we have selected the 60 percent confidence level and satisfaction of this ISO requirement means that a CMMI-rated organization would have to develop a problem management process. There are two ISO 20000 requirements that deserve further analysis, namely, service process improvement based on the reported problems and the reporting of the errors and their corrections to the incident management process. Service process improvement is fully covered

Table 5.23 Mapping of CAR SPs to ISO 20000

CMMI		*ISO 20000* Total	8.2 0	8.3 4	9.1 0
CAR	SP 1.1	0			
CAR	SP 1.2	1		X	
CAR	SP 2.1	1		X	
CAR	SP 2.2	1		X	
CAR	SP 2.3	1		X	

by CAR SP 1.2 and SP 2.3 and is further supported by OID SP 2.1 for planning the deployment of improvements. However, error and correction reporting has no equivalent in CMMI and will have to be specifically addressed to satisfy this ISO 20000 requirement (see the interfaces between those processes in chapter 4, Figure 4.6). CAR SP 1.1 Select Defect Data for Analysis is not mapped to ISO 20000, which means that an ISO registered organization that aims at CMMI ML/CL 5 will have to specifically address this process.

Following the same order of discussion as for ISO 9001, let's take a look at the OT PA, as shown in Table 5.24. Here we see a dramatic difference from the CMMI to ISO 9001 mapping shown in Table 5.4.

Table 5.24 Mapping of OT SPs to ISO 20000

CMMI		*ISO 20000* Total	3.1 0	3.2 0	3.3 2	4.1 0	4.2 0	4.3 0	4.4.1 0	4.4.2 0	4.4.3 0	5 3
OT	SP 1.1	2			X							X
OT	SP 1.2	1										X
OT	SP 1.3	1										X
OT	SP 1.4	0										
OT	SP 2.1	1			X							
OT	SP 2.2	0										
OT	SP 2.3	0										

What is this map telling us? Not all OT SPs are mapped to ISO 20000. ISO 20000 section 3.3 "Competence, awareness and training" and section 5 "Planning and implementing new or changed services" provide the maps. Section 3.3 provides an overall requirement for training, whereas section 5 deals with planning for training when a new or changed service is introduced. This is much less than CMMI suggests for developing a comprehensive organizational training capability that will include establishing training records and evaluating training effectiveness. Therefore, an ISO 20000 registered organization that would like to obtain a CMMI rating would have to specifically address the organizational training capability.

Let's now consider the Engineering PAs. As expected, due to the different focus of those two frameworks, most Engineering PAs are not mapped well. Again we will see markedly different maps from the ones between ISO 9001 and CMMI. Our first detailed map is between the RD and REQM PAs and ISO 20000 as shown in Table 5.25. Here we can see that several RD and REQM SPs were not mapped to ISO 20000 at all. In other cases, several SPs are mapped to the same ISO 20000 sections.

Table 5.25 Mapping of RD and REQM SPs to ISO 20000

		ISO 20000	3.2	4.1	5	6.1	6.3	6.4	6.5	6.6	7.2	7.3
CMMI		Total	3	1	4	9	3	1	2	2	3	2
RD	SP 1.1	2				X					X	
RD	SP 1.2	4	X			X	X				X	
RD	SP 2.1	4	X		X	X	X					
RD	SP 2.2	1					X					
RD	SP 2.3	2		X								X
RD	SP 3.1	0										
RD	SP 3.2	0										
RD	SP 3.3	2			X	X						
RD	SP 3.4	2			X	X						
RD	SP 3.5	1				X						
REQM	SP 1.1	3				X			X	X		
REQM	SP 1.2	2	X			X						
REQM	SP 1.3	7			X	X		X	X	X	X	X
REQM	SP 1.4	0										
REQM	SP 1.5	0										

What is this map telling us? In this table we can see that REQM and RD map to several common ISO 20000 requirements. The ISO 20000 section 6.4 "Budgeting and accounting for IT services" requirement is mapped only to a single REQM SP and the section 4.1 "Plan service management" requirement maps only to one RD SP. There are no maps between ISO 20000 and the following SPs:

- RD SP 3.1 Establish Operational Concepts and Scenarios
- RD SP 3.2 Establish a Definition of Required Functionality
- REQM SP 1.4 Maintain Bidirectional Traceability of Requirements
- REQM SP 1.5 Identify Inconsistencies Between Project Work Products and Requirements

What is the significance of those maps? Whereas the planning activities required by ISO 20000 section 4.1 "Plan service management" are covered in the PP and IPM PAs, the map to RD SP 2.3 addresses the interfaces between the service management processes. Those interfaces are also covered in the OPD and IPM PAs. The interfaces will require special attention during their documentation to ensure that they are planned. Because the RD and REQM maps are mostly at the medium (60 percent) level of confidence, those PAs will need to specifically address ISO 20000 requirements for documenting service level agreements; documenting changes to services; defining, recording and documenting service levels; reviewing and controlling those agreements; ensuring service continuity; and maintaining awareness of business needs.

As one might expect, the Technical Solution and Product Integration PAs are not mapped well, as shown in Table 5.26.

What is this map telling us? Although CMMI references to "products" are intended to include both product and services, CMMI practices require interpretation when used in the service environment. An example of interpretation of CMMI-DEV is found in chapter 6 in Chrissis et al. (2006), which describes a case study for applying CMMI to services at Raytheon's Pasadena Operations. If such an interpretation were used, there would be significantly more matches between TS and ISO 20000. Table 5.26 shows that both TS and PI map to ISO 20000 section 7.3 "Supplier management" and to no other requirements. Moreover, they only map to a requirement for documenting interfaces between service provider and acquirer to provide a seamless quality of service. There is virtually nothing in the ISO 20000 standard that deals with design, implementation, and integration of services.

On the other hand, VER and VAL are well represented as shown in Table 5.27.

Table 5.26 Mapping of TS and PI SPs to ISO 20000

		ISO 20000	*7.3*
CMMI		Total	3
TS	SP 1.1	0	
TS	SP 1.2	0	
TS	SP 2.1	0	
TS	SP 2.2	0	
TS	SP 2.3	1	X
TS	SP 2.4	0	
TS	SP 3.1	0	
TS	SP 3.2	0	
PI	SP 1.1	0	
PI	SP 1.2	0	
PI	SP 1.3	0	
PI	SP 2.1	1	X
PI	SP 2.2	1	X
PI	SP 3.1	0	
PI	SP 3.2	0	
PI	SP 3.3	0	
PI	SP 3.4	0	

What is this map telling us? It is not a surprise that VER SG 2 Perform Peer Reviews is not mapped to ISO 20000. In addition, the VER and VAL SPs dealing with selecting work products for verification and validation are also not mapped. Verification and validation planning as related to the new or changed services maps well to ISO, as does the requirement for establishing the verification or validation controlled acceptance test environment (ISO 20000 section 10.1). Verification specifically addresses the ISO 20000 requirements in sections 6.1 and 6.2 for monitoring service levels against the targets, identifying and reviewing nonconformances, and recording actions for improvement based on those findings. For a ISO 20000-registered organization that contemplates achieving a CMMI maturity level, this would mean that a more comprehensive view of verification and validation should be considered (e.g., peer reviews, technical reviews, and design reviews, in addition to testing).

As we did when analyzing the relationships between ISO 9001 and CMMI, we'll review how the CMMI Project Management PAs and ISO 20000 requirements

Table 5.27 Mapping of VER and VAL SPs to ISO 20000

CMMI		*ISO 20000* Total	*5* 6	*6.1* 1	*6.2* 1	*10.1* 2
VAL	SP 1.1	0				
VAL	SP 1.2	1				X
VAL	SP 1.3	1	X			
VAL	SP 2.1	1	X			
VAL	SP 2.2	1	X			
VER	SP 1.1	0				
VER	SP 1.2	1				X
VER	SP 1.3	1	X			
VER	SP 2.1	0				
VER	SP 2.2	0				
VER	SP 2.3	0				
VER	SP 3.1	1	X			
VER	SP 3.2	3	X	X	X	

relate. By inspecting the relationship matrix in Table 5.28, we can see that IPM simply supports PP and PMC and does not provide any additional maps to ISO 20000.

What is this map telling us? Here we see that the PP PA is mapped to many ISO 20000 requirements, whereas PMC may be satisfied with just one, namely, section 4.3 "Monitoring, measuring and reviewing (Check)."

IPM SP 1.2 Use Organizational Process Assets for Planning Project Activities, IPM SP 1.3 Establish Project's Work Environment, and IPM SP 1.6 Contribute to the Organizational Process Assets are not mapped at all. IPM SP 1.2 and SP 1.6 are consistent with PP SP 1.2 and their implementation may be accomplished at the same time. Without IPM SP 1.6, GP 3.2 in all PAs will be difficult to implement as can be seen in the example later dealing with the GPs.

The PMC SPs that map to section 4.3 "Monitoring, measuring and reviewing (Check)" show a 60 percent confidence level because ISO 20000 requirements for service management will have to be specifically addressed when developing and implementing CMMI-based plans. Service levels and noncompliances with established service level agreements required in ISO 20000 section 6.1 "Service level management" and section 6.2 "Service reporting" will have to be specifically measured and monitored when implementing CMMI PMC SG2 Manage Corrective Action

Table 5.28 Mapping of PP, PMC, and IPM SPs to ISO 20000

		ISO 20000	3.2	3.3	4.1	4.2	4.3	5	6.1	6.2	6.4	6.5	6.6	7.2	7.3
CMMI		Total	1	2	9	11	10	10	4	1	7	1	3	5	3
PP	SP 1.1	3			X			X			X				
PP	SP 1.2	0													
PP	SP 1.3	0													
PP	SP 1.4	2						X			X				
PP	SP 2.1	3			X			X			X				
PP	SP 2.2	2				X							X		
PP	SP 2.3	1	X												
PP	SP 2.4	3			X			X				X			
PP	SP 2.5	2		X				X							
PP	SP 2.6	2						X						X	
PP	SP 2.7	1			X										
PP	SP 3.1	1			X										
PP	SP 3.2	2						X			X				
PP	SP 3.3	3			X			X			X				
PMC	SP 1.1	4		X		X	X				X				
PMC	SP 1.2	1					X								
PMC	SP 1.3	3				X	X						X		
PMC	SP 1.4	1					X								
PMC	SP 1.5	1					X								
PMC	SP 1.6	2				X	X								
PMC	SP 1.7	4				X	X							X	X
PMC	SP 2.1	4					X		X	X			X		
PMC	SP 2.2	3					X		X						X
PMC	SP 2.3	2					X		X						
IPM	SP 1.1	3			X			X			X				
IPM	SP 1.2	0													
IPM	SP 1.3	0													
IPM	SP 1.4	1			X										
IPM	SP 1.5	1				X									
IPM	SP 1.6	0													
IPM	SP 2.1	2							X					X	
IPM	SP 2.2	1												X	
IPM	SP 2.3	2												X	X
IPM	SP 3.1	1				X									
IPM	SP 3.2	1				X									
IPM	SP 3.3	1				X									
IPM	SP 3.4	1				X									
IPM	SP 3.5	3			X	X		X							

to Closure. Capacity management, required by ISO 20000 section 6.5, can be successfully implemented by CMMI PP SP 2.4 Plan for Project Resources, provided that it is specifically addressed in that process. The situation is different when dealing with the requirements in ISO 20000 section 6.6 "Information security management," which are not addressed by CMMI, except when dealing with security risks. Therefore, we have mapped it to PP SP 2.2 Identify Project Risks and to PMC SP 1.3 Monitor Project Risks. This requirement also maps to all SPs in the Risk Management PA, but only at the 30 percent confidence level. The business relationship and supplier management requirements of ISO 20000 are addressed by PP SP 2.6, PMC SP 1.7, and SP 2.2 and are further supported by the Supplier Agreement Management PA (not shown in Table 5.28). The documentation and training requirements in ISO 20000 sections 3.2 and 3.3, respectively, may be addressed in plans for data management and training and then tracked through PMC SP 1.1.

An ISO 20000-registered organization would benefit from the CMMI processes for estimation (PP SP 1.2) and selection of an appropriate life cycle (PP SP 1.3) to scope the service project. Those two SPs do not map to ISO 20000 requirements.

We can now turn our attention to the Support PAs, beginning with the Measurement and Analysis PA, shown in Table 5.29.

What is this map telling us? In Table 5.29 we can see that ISO 20000 section 4.3 "Monitoring, measuring and reviewing (Check)" maps to all SPs in this PA. Here again we have the situation where the ISO 20000 standard requires monitoring, measuring, and controlling of the service management process. This specifically includes service levels (section 6.1), service availability (section 6.3), service capacity (section 6.5), and security incidents (section 6.6), which means that corresponding CMMI processes should address those topics as well as service management process improvement (ISO 20000 section 4.4.3).

Table 5.29 Mapping of MA SPs to ISO 20000

		ISO 20000	4.3	4.4.3	6.1	6.2	6.3	6.5	6.6	8.3
CMMI		Total	8	2	2	3	4	5	3	1
MA	SP 1.1	3	X					X	X	
MA	SP 1.2	4	X				X	X	X	
MA	SP 1.3	2	X					X		
MA	SP 1.4	4	X			X		X	X	
MA	SP 2.1	4	X	X			X			X
MA	SP 2.2	6	X	X	X	X	X	X		
MA	SP 2.3	1	X							
MA	SP 2.4	4	X		X	X	X			

Table 5.30 Mapping of CM and PPQA SPs to ISO 20000

		ISO 20000	3.2	4.3	4.4.1	5	6.3	6.4	6.6	7.2	7.3	8.3	9.1	9.2	10.1
CMMI		Total	1	2	1	2	2	1	1	2	2	2	6	3	2
CM	SP 1.1	2											X	X	
CM	SP 1.2	1											X		
CM	SP 1.3	1											X		
CM	SP 2.1	8				X	X	X		X	X	X		X	X
CM	SP 2.2	9				X	X		X	X	X	X	X	X	X
CM	SP 3.1	1											X		
CM	SP 3.2	1											X		
PPQA	SP 1.1	0													
PPQA	SP 1.2	0													
PPQA	SP 2.1	2		X	X										
PPQA	SP 2.2	2	X	X											

Table 5.30 shows the mapping of CM and PPQA process areas to ISO 20000.

What is this map telling us? In Table 5.30 we can see that CM SPs are well mapped to ISO 20000, whereas PPQA SPs are very sparsely mapped. CM is one of the most important support processes in the ISO 20000 standard. It is specifically called for in section 9.1 "Configuration management" and section 9.2 "Change management," which together map to all CM SPs addressing the requirements for service continuity, availability, and capacity management, among others. However, even though there is a strong match, there are several important ISO 20000 requirements that are not mapped to CMMI, such as:

- Classifying change requests as urgent, emergency, major, minor; reversing changes if proved unsuccessful (a very interesting requirement that would benefit many CMMI-rated organizations)
- Authorization of emergency changes (normally planned in most CMMI-rated organizations, but not specifically called for by CMMI)
- Analyzing changes and drawing conclusions from those analyses

Some of the more interesting aspects of ISO 20000 requirements that can be implemented by CM SP 2.1 Track Change Requests and SP 2.2 Control Configuration Items are the requirements for evaluating the impact of potential changes on the availability and continuity of service, changes that are required to correct cases of the problems, and several aspects of release management. Those requirements will have to be specially addressed in the CMMI CM process. Going in the opposite direction, implementing the CM PA will satisfy not only the Control processes requirements but also some requirements of section 10.1 "Release management process."

ISO 20000 has no requirement for establishing a process for objective evaluation of processes and work products (PPQA SG 1). ISO 20000 only requires audit planning, communicating noncompliance issues, establishing records, and reporting noncompliances. An ISO registered organization may benefit from implementing the CMMI practices for objective evaluation. Audit plans may, for example, contain those additional processes. Note that the ISO requirements for managing, auditing, and improving quality of service is mapped to PPQA GP 2.2, which is not shown in Table 5.30.

The CMMI organizational processes are not well represented in ISO 20000 as shown in Table 5.31.

Table 5.31 Mapping of OPF and OPD SPs to ISO 20000

		ISO 20000	*3.2*	*4.4.2*	*4.4.3*	*7.3*	*9.2*	*10.1*
CMMI		Total	1	7	8	1	2	1
OPD	SP 1.1	5	X	X	X	X		X
OPD	SP 1.2	0						
OPD	SP 1.3	0						
OPD	SP 1.4	0						
OPD	SP 1.5	0						
OPD	SP 1.6	0						
OPD	SP 2.1	0						
OPD	SP 2.2	0						
OPD	SP 2.3	0						
OPF	SP 1.1	1			X			
OPF	SP 1.2	2		X	X			
OPF	SP 1.3	3		X	X		X	
OPF	SP 2.1	3		X	X		X	
OPF	SP 2.2	2		X	X			
OPF	SP 3.1	2		X	X			
OPF	SP 3.2	2		X	X			
OPF	SP 3.3	0						
OPF	SP 3.4	0						

What is this map telling us? The ISO 20000 section 3.2 requirement for documenting processes and procedures maps well to OPD SP 1.1. There are several other ISO 20000 requirements that map to that SP, such as:

- Section 4.4.2 Process for continual process improvement
- Section 4.4.3 Process for revision of policies, procedures, and plans
- Section 7.3 Process for supplier management
- Section 10.1 Release management process

However, no other SPs in OPD are mapped, so an ISO 20000 organization will have to address those practices to achieve CMMI CL/ML 3. Regardless of ML/CL achievement, an organization that contemplates implementing ISO 20000 would benefit from implementing many, if not all, of the OPD practices.

On the other hand, the process audit aspects of ISO 20000 section 4.4 and some aspects of process asset deployment (sections 4.4.2 "Management of improvement," 4.4.3 "Activities," and 10.1 "Release management process") are well represented in OPF. There are strong ISO maps to OPF SG 1 Determine Improvement Opportunities and SG 2 Plan and Implement Process Improvements. Some aspects of process deployment addressed by OPF SG 3 such as monitoring implementation and assessing process effectiveness are not mapped. Implementing those practices may benefit ISO 20000 institutionalization.

Organizational high maturity PAs are shown in Table 5.32.

Table 5.32 Mapping of OPP and OID SPs to ISO 20000

		ISO 20000	*4.4.2*	*4.4.3*	*7.2*	*8.3*
CMMI		Total	6	8	1	1
OPP	SP 1.1	0				
OPP	SP 1.2	0				
OPP	SP 1.3	2		X	X	
OPP	SP 1.4	0				
OPP	SP 1.5	0				
OID	SP 1.1	2	X	X		
OID	SP 1.2	2	X	X		
OID	SP 1.3	1		X		
OID	SP 1.4	2	X	X		
OID	SP 2.1	3	X	X		X
OID	SP 2.2	2	X	X		
OID	SP 2.3	2	X	X		

What is this map telling us? By comparing the mapping in Table 5.31 to Table 5.32, we can see that OID and OPF support the same ISO 20000 requirements. This is interesting since OID is an advanced PA that builds on the processes established in the OPF PA. The exception is the ISO 20000 section 8.3 requirement addressing problem management and the actions that can be used to further improve the service. On the other hand, only one SP in OPP (SP 1.3 Establish Quality and Process Performance Objectives) is mapped to ISO 20000. This SP maps to requirements for service providers to set targets for improving quality, costs, and resource utilization (section 4.4.3) and to be aware of business needs (section 7.2). The quantitative aspects of managing quality and process performance, as required by the QPM PA, are only addressed in ISO 20000 section 6.5 "Capacity management," and specifically in requiring predictive analysis of business needs. There is very little in the standard that requires quantitative management but many service management processes can be managed quantitatively. For example, using statistical techniques in continuity and availability management (section 6.3), capacity management (section 6.5), incident management (section 8.2), problem management (section 8.3), and control processes (section 9) may provide great insight for service improvement.

Let's now revisit the GPs. The DAR, OPP, OT, PI, QPM, and TS PAs have no GP maps to ISO 20000, whereas OID and OPD have very sparse maps. In addition, the ISO 20000 requirements in section 6.3 "Service continuity and availability management," section 6.5 "Capacity management," section 7.2 "Business relationship management," and section 8.2 "Incident management" did not map to any GPs.

By inspecting the overall CMMI-to-ISO 20000 mapping, one can see that some GPs did not map to ISO, as shown in Table 5.33.

Table 5.33 GP to ISO 20000 Mapping Example

CMMI		*ISO 20000 Matches*
PMC	GP 2.1	X
PMC	GP 2.2	X
PMC	GP 2.3	X
PMC	GP 2.4	X
PMC	GP 2.5	X
PMC	GP 2.6	0
PMC	GP 2.7	0
PMC	GP 2.8	X
PMC	GP 2.9	X
PMC	GP 2.10	X
PMC	GP 3.1	X
PMC	GP 3.2	0

What is this map telling us? By inspecting the matrices we can see that most GPs map to ISO 20000. If we examine the PMC mapping as a typical example, we can see that there is no mapping to GP 2.6, 2.7, and 3.2. This implies that an ISO-registered organization would be, at best, rated CL 1 on those PAs that do not have GP 2.6 or GP 2.7 implemented, and CL 2 for those PAs that do have mappings for all GP 2.x since none have GP 3.2 mapped.

This raises concerns about institutionalization and the role GPs have in establishing and maintaining processes that are used to provide, control, and manage services. It is recommended that an organization targeting IT process improvement and ISO 20000 registration would implement those generic practices to provide a stable infrastructure for process improvement and implementation.

If we look at the results of the ISO 20000-to-CMMI mapping rolled up to the ISO section/subsection level and to CMMI PA level, as shown in Table 5.34, we see that only ISO section 8.2 "Incident management" has no mapping at all, whereas sections 6.6 "Information security management" and 10.1 "Release management process" have sparse maps.

The summary result of the CMMI-to-ISO 20000 (inverse) map is shown in Table 5.35. In this table we see that DAR is not mapped at all; PI and TS are mapped with one or two practices; and OPD, OT, and QPM are mapped very weakly.

What is this map telling us? There are several points that can be made about the rolled-up results. The first and foremost concerns maintaining focus on the process improvement goal. An organization that has a goal of improving service management processes will greatly benefit from the process institutionalization and rigor provided by CMMI. CMMI can guide an organization in establishing and maintaining a strong process infrastructure while addressing the specific service management requirements outlined in ISO 20000. Intuitively, such an organization will be better positioned to achieve ISO 20000 registration and will find it much easier to maintain this registration over time. An organization that is CMMI-rated will need to add service management processes to its organizational standard. By using the continuous representation, efforts can be further streamlined by selecting those PAs and CLs needed to satisfy ISO 20000, which will provide the greatest return on investment through institutionalization.

Translating the mapping results to maturity levels, it appears that an organization rated at maturity level 3 would be able to satisfy most of the ISO 20000 requirements providing that it addresses services, service management, and service level agreements when developing and documenting its processes. The organization will have to specifically develop processes for incident management, which has no corresponding process in CMMI. By implementing selected elements of CAR and OID, ISO 20000 could be fully satisfied. Moreover, CAR can be implemented as

Table 5.34 CMMI–ISO 20000 Mapping Summary

ISO 20000	*CMMI*
Section 3	
Management responsibility Documentation requirements Training	CM, OPD, OT, PMC, PP, RD, REQM, RSKM, SAM, VAL, VER
Section 4	
Planning, implementing, monitoring, and continually improving service management	CM, IPM, MA, OID, OPD, OPF, PMC, PP, PPQA, RD, REQM, RSKM, SAM, VAL, VER
Section 5	
Planning and implementing new or changed services	CM, MA, OT, PMC, PP, RD, REQM, RSKM, VAL, VER
Section 6	
6.1: Service level management	MA, PMC, REQM, VER
6.2: Service reporting	MA, VER
6.3: Service continuity and availability management	MA
6.4: Budgeting and accounting for IT services	PMC, PP, REQM, RSKM
6.5: Capacity management	MA, PP, REQM
6.6: Information security management	MA
Section 7	
7.1: General	No requirements
7.2: Business relationship management	CM, IPM, PMC, RD, SAM
7.3: Supplier management	CM, OPD, PMC, RD, REQM, SAM
Section 8	
8.1: Background	No requirements
8.2: Incident management	None
8.3: Problem management	CAR
Section 9	
9.1: Configuration management	CM
9.2: Change management	CM
Section 10	
10.1: Release management process	CM

Table 5.35 Results of CMMI-ISO 20000 Mapping

Very Weak	*Not Mapped*
OPD	DAR
OT	OPP (1 practice)
QPM	PI (2 practices)
	TS (1 practice)

a problem-solving process without quantitative process knowledge, whereas OID should address continual process improvement in a manner that extends the OPF processes.

The situation is markedly different for those organizations that are ISO 20000 registered and would like to achieve a CMMI ML 3 rating. Since two PAs that characterize ML 3, OPD and OT, are sparsely mapped, such an organization will first need to reevaluate its standard processes, including those related to service management; revisit its organizational training capability; and interpret the DAR, TS, and PI PAs for service management. Moreover, since GP 2.5, 2.6, 2.7, and 3.2 are seldom mapped, such an organization would need an extra effort to rise above CL 1. Therefore, special attention will be needed for institutionalization of the PAs associated with those GPs. These conclusions are consistent with the architecture of the draft CMMI-SVC.

Table 5.36 summarizes ISO 20000 requirements that require special attention.

Table 5.36 Summary of the ISO 20000 Requirements That Require Special Attention

ISO Requirement	*Description*
3.0 Requirements for a Management System	
3.1—Evidence of top/executive management commitment	An organization has to provide evidence that its top management is committed to developing, implementing, and improving its SM capability. However, for the SM organizations, such management actions must be planned based on the business objectives and customer requirements.
3.1—Management responsibility: Establish service management mandates and guidelines	An organization has to establish the SM policy, objectives, and plans. (In CMMI, there is no requirement for such policy, objectives, and plans, but such policies and plans can be written and implemented using GP 2.1, GP 2.2 in PP, PMC process areas.)

ISO Requirement	*Description*
3.1—Management responsibility: Communicate service management priorities	An organization has to communicate the importance of achieving SM objectives and its continual improvement (there is no requirement in CMMI for communicating SM objectives). If SM is properly implemented, this requirement will be satisfied. It may be incorporated in OPF SP 2.1.
3.1—Management responsibility: Allocate resources	In CMMI, there is no requirement for management to provide resources for SM. However, this can be achieved by implementing GP 2.3 in appropriate PAs.
3.2—Service management documentation: Additional required records	In general, in CMMI there are no requirements to collect records. Records (objective evidence) are associated with appraisals. This requirement is implicit in CMMI, but a special process can be written for that purpose.
3.3—Maintain employee awareness	There is no corresponding requirement in CMMI.
4.0 Planning and Implementing Service Management	
4.1—Service management plan: Project interfaces	In CMMI, interfaces among projects is not addressed.
4.3—Conduct management reviews: Review conformance with plan and requirements	The ISO text is "review requirements of this standard," which maps to PMC SP 1.6 and SP 1.7, but those reviews will also have to address ISO requirements.
5.0 Planning and Implementing New or Changed Services	
5—Outcomes reporting and postimplementation review	Not in CMMI (it is postimplementation).
6.0 Service Delivery Process	
6.2—Service reporting: Satisfaction analysis	CMMI does not sufficiently cover "satisfaction." In CMMI, service continuity and availability are not addressed.
6.4—Report costs	"Financial forecast" is not addressed in PP/PMC.
6.5—Capacity management	Capacity Management is not covered by CMMI but it can be interpreted (specifically PP, REQM, and MA).
6.6—Information security	In CMMI, security is not addressed. Some security aspects could be implemented by interpreting the existing CMMI practices, but in most cases this may not be adequate because security has to be built into the processes and not simply added as an afterthought.

(continued)

Table 5.36 Summary of the ISO 20000 Requirements That Require Special Attention (continued)

ISO Requirement	*Description*
7.0 Relationship Processes	
7.2—Establish complaints process Establish formal service complaint definition Resolve formal service complaints Provide customer complaint escalation process	In CMMI, complaints and the complaint process are not addressed.
7.2—Establish process for responding to customer satisfaction measurements	In CMMI, "customer satisfaction" is not addressed.
7.2—Identify improvement actions based on service complaints	In CMMI "improvement actions based on service complaints" are not addressed.
7.3—Ensure supplier SLAs are aligned with the SLAs with the business	Not in CMMI. However, it can be implemented by providing required alignment.
7.3—Document supplier agreement and obtain agreement(s)	"Level of service" should be addressed.
7.3—Establish contractual disputes process	This ISO requirement is not addressed in CMMI.
8.0 Resolution Processes	
8.2—Record incidents	Incident management is a postdelivery process and is not covered in CMMI.
9.0 Control Processes	
9.1—Establish interface to financial asset accounting processes	The CMMI is silent on this topic.
9.2—Document and classify requests for change	In CMMI there are no such guidelines but CM processes usually specify this.
9.2—Reverse/remedy unsuccessful change	Although this would be part of adequate planning for change implementation, it is not addressed in CMMI.

ISO Requirement	*Description*
9.2—Establish emergency change policies and procedures	Although it could be handled as a part of CM GP 2.1 and 2.2, it is not explicitly required in the CMMI.
9.2—Analyze change records	In CMMI, this is not explicitly called out as it relates to CM for services.
9.2—Record results of change record analyses	CMMI is silent in this area.
10.0 Release Process	
10.1—Document release policy	Although there is no direct CMMI requirement to establish a release policy, it could certainly be covered by CM GP 2.1 or 2.2 (see General statements).
10.1—Reverse/remedy unsuccessful release	There is nothing in CMMI that addresses reversing the release if not successful. This will have to be described in the CM process.
10.1—Develop the plan for release of services, systems, software, and hardware	CMMI does not enumerate all the requirements listed in ISO therefore the CM Plan (GP 2.2) should address all of those requirements.
10.1—Release management process	CMMI does not address Incident Management.
10.1—Establish process for managing emergency releases	The CMMI is silent on special provisions for emergency releases.
10.1—Maintain integrity of hardware and software during release and distribution	CM SG3 addresses integrity of the CM baseline but does not mention anything about integrity during installation, handling, packaging, and delivery directly.
10.1—Measure release successes and failures Include postrelease period in measures Analyze release incidents Identify improvement actions based on release management process	There are no specific CMMI requirements. (This is a postdelivery activity.)

ISO 15288:2008 to CMMI Maps

As noted earlier, ISO 15228 and ISO 12207 are life cycle management standards and provide much better matches to CMMI than ISO 9001 or ISO 20000. However, both ISO 12207 and ISO 15288 rely on ISO 15504 for process institutionalization, meaning that CMMI GPs will be lightly mapped. Our discussion of these standards will concentrate on their contribution to process improvement rather than addressing their role as Process Reference Models (PRMs) in assessments using ISO 15504.

As indicated in chapter 4, the standard defines the purpose, outcomes, and activities and tasks for each process. Our mapping concentrates on the activities and tasks. The complete ISO 15288–CMMI map is given in Appendix E. There are no CMMI PAs that were not mapped to ISO 15288, although some were mapped more sparsely than others. For example, high maturity PAs, such as OPP and QPM, are not fully mapped to ISO 15288. Going in the other direction, from ISO 15288 to CMMI, sections 6.4.9 "Operations Process," 6.4.10 "Maintenance Process," and 6.4.11 "Disposal Process" are not mapped since those subjects are not covered by CMMI.

We learn the following from these mappings:

- A number of PAs have strong maps to one or two ISO 15288 sections. These are shown in Table 5.37.
- Other PAs, such as PPQA, OPD, OPF, OT, OID, and OPP, have mappings that are spread over a larger number of ISO 15288 sections.
- Some SPs do not map to any ISO 15288 requirements. Some of the more significant SPs in this category are shown in Table 5.38.

This means that for those practices that do not map, CMMI-based process improvement will not benefit from ISO 15288 and that an ISO 15288-compliant organization would need to specifically address those unmapped CMMI practices.

ISO 15288 relies on ISO 15504 for the description of institutionalization and the generic practices in CMMI make institutionalization an integral part of the model. Thus, the GPs usually do not have strong maps. In particular, GP 2.5, GP 2.6, GP 3.1, and GP 3.2 are infrequently mapped to ISO requirements. This has a very strong negative effect on the institutionalization of ML 2 and ML 3 processes if an ISO-compliant organization attempts to implement CMMI. Such an organization would, in addition to implementing ISO 15288 practices, need to consider Annex B of ISO 12207 where process attributes similar to CMMI GPs are addressed. However, as we will see in the examples below, some ISO 15288 requirements are mapped to GPs, indicating that some aspects of institutionalization, as defined by CMMI, are implicit in the ISO standard.

The ISO 15288 Annex A discussion of tailoring differs from the CMMI tailoring concepts. CMMI talks about tailoring an organizational set of standard processes,

Table 5.37 CMMI PAs Closely Mapped to ISO 15288 Sections

PA	*ISO 15288 Section*
DAR	6.3.3 Decision Management Process
RSKM	6.3.4 Risk Management Process
CM	6.3.5 Configuration Management Process
MA	6.3.7 Measurement Process
RD, REQM	6.4.1 Stakeholder Requirements Definition Process and 6.4.2 Requirements Analysis Process
PP	6.3.1 Project Planning Process (with some IPM overlap)
PMC	6.3.2 Project Assessment and Control Process (with some IPM, REQM, and SAM overlap)
PP, PMC	6.3.6 Information Management Process
TS	6.4.3 Architectural Design Process (with some RD overlap) and 6.4.4 Implementation Process
PI	6.4.5 Integration Process and 6.4.7 Transition Process
VER	6.4.6 Verification Process (not all SPs are mapped but all ISO 15288 "shall" statement were satisfied)
VAL	6.4.8 Validation Process

Table 5.38 CMMI Practices That Do Not Map to ISO 15288

PA	*SP*	*Title*
IPM	1.6	Contribute to the Organizational Process Assets
IPM	2.1	Manage Stakeholder Involvement
OPD	1.4	Establish Organization's Measurement Repository
OPD	1.5	Establish Organization's Process Asset Library
OPF	3.1	Deploy Organizational Process Assets
OPF	3.2	Deploy Standard Processes
OT	1.2	Determine Which Training Needs Are the Responsibility of the Organization
OT	2.3	Assess Training Effectiveness
PI	1.3	Establish Product Integration Procedures and Criteria
PI	2.1	Review Interface Descriptions for Completeness
REQM	1.1	Obtain Understanding of Requirements
RSKM	1.3	Establish Risk Management Strategy
SAM	2.5	Transition Products
TS	3.2	Develop Product Support Documentation
VER	2.1	Prepare for Peer Reviews
VER	2.3	Analyze Peer Review Data

whereas Annex A discusses tailoring of the standard. This would be equivalent to tailoring CMMI for a specific organization or project purpose.

Before we start analyzing the mapping examples, let's explain the notation used in our examples. Each life cycle process discussed in the standard is structured to show the process name (section 6.x.y), purpose (6.x.y.1), outcomes (6.x.y.2), and activities and tasks (6.x.y.3). Individual tasks are numbered 6.x.y.3 a1, a2, . . . , b1, b2, . . . , and so forth. Since our mapping is performed at the activity/task shall level, only sections 6.x.y.3, and their individual requirements a1, a2, . . . , and so forth are indicated in the examples. In several cases, an ISO 15288 requirement references requirements in several other sections. For example, the section 6.3.4.3 "Risk Management Process" requirement, references section 6.3.2.3 "Project Assessment and Control Process" and, in turn, its mapping to CMMI. In general, we do not provide additional mapping information for such cases.

Let's now take a closer look at some of the ISO 15288 sections that mapped well to CMMI PAs. The section 6.3.3 "Decision Management Process" mapping is shown in Table 5.39. Here we can see that both the DAR PA and the section

Table 5.39 Mapping DAR to ISO 15288 Section 6.3.3

		ISO 15288	*6.3.3.3*							
			a1	*a2*	*a3*	*b1*	*b2*	*b3*	*c1*	*c2*
CMMI		Total	2	3	1	1	1	1	1	2
DAR	SP 1.1	2	X	X						
DAR	SP 1.2	1		X						
DAR	SP 1.3	1		X						
DAR	SP 1.4	2				X	X			
DAR	SP 1.5	1						X		
DAR	SP 1.6	2							X	X
DAR	GP 2.1	0								
DAR	GP 2.2	1	X							
DAR	GP 2.3	0								
DAR	GP 2.4	0								
DAR	GP 2.5	0								
DAR	GP 2.6	1								X
DAR	GP 2.7	1			X					
DAR	GP 2.8	0								
DAR	GP 2.9	0								
DAR	GP 2.10	0								
DAR	GP 3.1	0								
DAR	GP 3.2	0								

6.3.3 requirements are completely satisfied. The expected outcomes for this section match the typical DAR work products. Also note that requirement 6.3.3.3 a3 is not mapped to an SP and requires implementation of GP 2.7. However, some important GPs are not mapped, including GP 2.5 Train People, GP 2.9 Objectively Evaluate Adherence, and GP 3.1 Establish a Defined Process.

The section 6.3.4 "Risk Management Process" mapping is shown in Table 5.40. As noted in Table 5.38, RSKM SP 1.3 does not map to ISO 15288 although it is required by outcome 6.3.4.2 b. Documentation of the risk management strategy is part of risk management planning (requirement a2) and will completely satisfy CMMI. Several requirements are mapped to the GPs, thereby providing some institutionalization opportunities. Notes in this section may provide additional information for a CMMI-based risk management process.

In the next example, we see that the CM PA maps fully to section 6.3.5 "Configuration Management Process," as shown in Table 5.41. Here, too, several

Table 5.40 Mapping RSKM to ISO 15288 Section 6.3.4

		ISO 15288	*6.3.4.3*																					
			a1	*a2*	*a3*	*a4*	*a5*	*b1*	*b2*	*b3*	*b4*	*c1*	*c2*	*c3*	*c4*	*d1*	*d2*	*d3*	*e1*	*e2*	*e3*	*f1*	*f2*	*f3*
CMMI		Total	1	2	2	1	2	1	1	1	1	1	1	1	1	1	1	1	1	1	1	1	1	2
RSKM	SP 1.1	1						X																
RSKM	SP 1.2	1							X															
RSKM	SP 1.3	0																						
RSKM	SP 2.1	2										X									X			
RSKM	SP 2.2	3								X			X	X										
RSKM	SP 3.1	2													X	X								
RSKM	SP 3.2	3															X	X	X					
RSKM	GP 2.1	1	X																					
RSKM	GP 2.2	1		X																				
RSKM	GP 2.3	1				X																		
RSKM	GP 2.4	1			X																			
RSKM	GP 2.5	0																						
RSKM	GP 2.6	0																						
RSKM	GP 2.7	2			X						X													
RSKM	GP 2.8	3					X													X			X	
RSKM	GP 2.9	0																						
RSKM	GP 2.10	1																						X
RSKM	GP 3.1	1		X																				
RSKM	GP 3.2	3					X															X		X

Table 5.41 Mapping CM to ISO 15288 Section 6.3.5

		ISO 15288	*6.3.5.3*			
			a1	*a2*	*b1*	*b2*
CMMI		Total	1	1	3	4
CM	SP 1.1	1		X		
CM	SP 1.2	1	X			
CM	SP 1.3	1			X	
CM	SP 2.1	2			X	X
CM	SP 2.2	1				X
CM	SP 3.1	2			X	X
CM	SP 3.2	1				X

ISO 15288 notes provide additional information that may be helpful to a CMMI-based CM process.

The measurement process was added to the 2008 version of the standard and it maps completely to the MA PA as shown in Table 5.42. MA SP 1.2 is also mapped to section 6.4.2.3 a4, which deals with the definition of technical and "quality in use" measurements such as satisfaction, reliability, and safety as defined in ISO 9126-4:2004 (*Software Engineering – Product quality – Part 4: Quality in use metrics*). Note that most MA GPs are not mapped to ISO 15288.

Let's now take a look at the Engineering PAs. RD and REQM are shown separately from TS–PI and VER–VAL since they exhibit interesting mapping properties. RD and REQM map to sections 6.4.1 "Stakeholder Requirements Definition Process" 6.4.2 "Requirements Analysis Process," and 6.4.3 "Architectural Design Process" as shown in Table 5.43.

What is this map telling us? Here we see that the RD PA satisfies most of the section 6.4 requirements, whereas PP provides the mapping to the requirement that deals with stakeholder identification, and REQM SP 1.4 provides the mapping that deals with maintaining traceability to stakeholder needs and between requirements. REQM SP 1.3 supports recording requirements in conjunction with RD SP 1.2 and SP 2.2, which support developing and allocating requirements. Extensive notes in this section cross-reference several ISO standards including ISO 9126-1:2001 (*Software engineering – Product quality – Quality model*) and ISO 13407:1999 (*Ergonomics – Ergonomics in human-systems interaction – Human-centered design process for interactive systems*). However, since REQM SP 1.1, SP 1.2, and SP 1.5 are not mapped, ISO 15288 does not provide additional support to CMMI process improvement.

Table 5.42 Mapping MA to ISO 15288 Section 6.3.7

		ISO 15288	*6.3.7.3*													*6.4.2.3 a4*
			a1	*a2*	*a3*	*a4*	*a5*	*a6*	*a7*	*b1*	*b2*	*b3*	*b4*	*c1*	*c2*	
CMMI		Total	1	1	1	1	1	1	1	1	2	1	1	1	1	1
MA	SP 1.1	2	X	X												
MA	SP 1.2	2			X											X
MA	SP 1.3	1				X										
MA	SP 1.4	1					X									
MA	SP 2.1	1									X					
MA	SP 2.2	1										X				
MA	SP 2.3	1									X					
MA	SP 2.4	1											X			
MA	GP 2.1	0														
MA	GP 2.2	1								X						
MA	GP 2.3	2						X	X							
MA	GP 2.4	0														
MA	GP 2.5	0														
MA	GP 2.6	0														
MA	GP 2.7	0														
MA	GP 2.8	0														
MA	GP 2.9	1												X		
MA	GP 2.10	0														
MA	GP 3.1	0														
MA	GP 3.2	1													X	

The TS and PI PA maps are shown in Tables 5.44A and 5.44B.

What are these maps telling us? We see that TS maps mostly to section 6.4.3 “Architectural Design Process” and section 6.4.4 “Implementation Process” as expected. We also see interaction of requirements 6.2.3.3 a1 and a3, which deal with logical architecture design and interfaces at the system boundary, respectively, with RD, as shown in Table 5.43. One TS practice, TS SP 3.2 Develop Product Support Documentation, did not map at all. The topics covered by both frameworks are quite similar with additional explanations found in the ISO 15288 notes. These notes, which are associated with specific ISO requirements, are at times quite detailed and may help CMMI-based process improvement by providing helpful implementation hints.

In the case of the PI PA, most of the mapping comes from sections 6.4.5 “Integration Process” and 6.4.7 “Transition Process.” PI SP 1.3 Establish Product

Table 5.43 Mapping RD and REQM to ISO 15288 Section 6.4

	ISO 15288	6.4.1.3												6.4.2.3									6.4.3.3				
		a1	a2	b1	b2	b3	b4	c1	c2	c3	c4	c5	c6	a1	a2	a3	a4	a5	b1	b2	b3	b4	a1	a2	a3	b2	c3
CMMI	Total	1	1	2	1	1	5	1	2	1	1	3	1	2	1	2	1	5	2	1	1	2	2	2	1	2	1
PP SP 2.6	1	X																									
RD SP 1.1	4		X	X			X									X											
RD SP 1.2	5			X			X					X						X				X					
RD SP 2.1	3						X											X				X					
RD SP 2.2	5											X				X							X	X		X	
RD SP 2.3	4						X							X											X	X	
RD SP 3.1	3				X	X								X													
RD SP 3.2	4						X								X								X	X			
RD SP 3.3	5							X	X								X	X	X								
RD SP 3.4	2								X									X									
RD SP 3.5	5									X	X							X	X	X							
REQM SP 1.1	0																										
REQM SP 1.2	0																										
REQM SP 1.3	1											X															
REQM SP 1.4	3												X								X						X
REQM SP 1.5	0																										

Table 5.44A Mapping TS to ISO 15288

		ISO 15288	*6.4.2.3 a3*	*6.4.3.3*							*6.4.4.3*			*6.4.6.3 a3*
				a1	*a3*	*b1*	*b3*	*b4*	*c1*	*c2*	*a2*	*b1*	*b2*	
CMMI		Total	1	2	1	2	3	2	2	2	1	1	1	1
TS	SP 1.1	4		X		X	X	X						
TS	SP 1.2	5	X	X			X	X	X					
TS	SP 2.1	4				X			X	X				X
TS	SP 2.2	1								X				
TS	SP 2.3	1			X									
TS	SP 2.4	1					X							
TS	SP 3.1	3									X	X	X	
TS	SP 3.2	0												

Table 5.44B Mapping PI to ISO 15288

		ISO 15288	*6.4.4.3 b3*	*6.4.5.3*							*6.4.7.3*						
				a1	*a2*	*b1*	*b2*	*b3*	*b4*	*b5*	*a1*	*a2*	*b1*	*b2*	*b3*	*b4*	*b5*
CMMI		Total	1	1	1	1	1	1	2	1	1	1	1	1	1	1	1
PI	SP 1.1	2		X	X												
PI	SP 1.2	1				X											
PI	SP 1.3	0															
PI	SP 2.1	0															
PI	SP 2.2	1							X								
PI	SP 3.1	2					X	X									
PI	SP 3.2	1							X								
PI	SP 3.3	1								X							
PI	SP 3.4	8	X								X	X	X	X	X	X	X

Integration Procedures and Criteria, and SP 2.1 Review Interface Descriptions for Completeness are not mapped and will have to be addressed separately. In this last case, it may be assumed that requirement 6.4.5.3 b3, which deals with verification and validation of components, would address component interfaces. Section 6.4.5 also includes ample notes that clarify requirements and may assist CMMI process developers.

Now let's take a look at the VER and VAL PAs, as shown in Table 5.45.

Table 5.45 Mapping VAL and VER to ISO 15288

		ISO 15288	*6.4.4.3 b2*	*6.4.6.3*						*6.4.7.3 b5*	*6.4.8.3*						
				a1	*a2*	*b1*	*b2*	*b3*	*b4*		*a1*	*a2*	*b1*	*b2*	*b3*	*b4*	*b5*
CMMI		Total	1	3	1	1	2	1	1	1	2	1	1	1	1	2	1
VAL	SP 1.1	1									X						
VAL	SP 1.2	2									X		X				
VAL	SP 1.3	1										X					
VAL	SP 2.1	2												X		X	
VAL	SP 2.2	4								X					X	X	X
VER	SP 1.1	1		X													
VER	SP 1.2	2		X		X											
VER	SP 1.3	2		X	X												
VER	SP 2.1	0															
VER	SP 2.2	1					X										
VER	SP 2.3	0															
VER	SP 3.1	1					X										
VER	SP 3.2	3	X					X	X								

What is this map telling us? As we can see, there is no overlap between the maps for these two PAs. The VAL PA maps fully to section 6.4.8 "Validation." The VER mapping to section 6.4.3.3, however, does not include peer review preparation and data analysis, which will have to be specifically addressed. Peer reviews in ISO 15288 are limited to design only (6.4.6.3 b2). Needless to say, CMMI is much more specific in peer review expectations than ISO. The VAL SP 2.2 and VER SP 3.2 practices dealing with analysis of results are also mapped to ISO 15288 requirements for system demonstration (6.4.7.3 b5) and recording evidence that agreements and policies (6.4.4.3 b2) are satisfied.

The CMMI mapping to section 6.2.5 "Quality Management Process" is shown in Table 5.46.

What is this map telling us? The Quality Management Process mapping is not "monolithic" as we saw for the DAR and MA PAs. Rather, it maps to several PAs. In addition to the SPs, there is significant mapping to the GPs from several PAs, summarized at the bottom of the table. Specifically GP 2.9, which addresses the objective evaluation of work products, and GP 2.10, which deals with senior management review, both map to section 6.2.5.3 b2 "Plan reviews." GP 2.4, which requires definition of authority and responsibility, maps to section 6.2.5.3 a3 "Define responsibility." In addition, section 6.3.2 "Project Assessment and Control Process" also addresses quality assurance and maps to all PPQA SPs. We also see

Table 5.46 CMMI Mapping to ISO 15288 Quality Management Process

		ISO 15288	*6.2.5.3*								*6.3.2.3*	
			a1	*a2*	*a3*	*b1*	*b2*	*b3*	*c1*	*c2*	*a2*	*b4*
CMMI		Total	1	4	1	2	4	4	1	1	4	1
PP	SP 3.1	1					X					
IPM	SP 1.4	1					X					
MA	SP 1.1	1		X								
MA	SP 2.2	1				X						
MA	SP 2.4	1				X						
PPQA	SP 1.1	1									X	
PPQA	SP 1.2	1									X	
PPQA	SP 2.1	3							X		X	X
PPQA	SP 2.2	2								X	X	
OID	SP 2.3	1						X				
OPD	SP 1.1	1	X									
OPF	SP 1.1	1		X								
OPF	SP 2.2	1						X				
OPP	SP 1.3	1		X								
QPM	SP 1.1	1		X								
QPM	SP 2.3	1						X				
	GP 2.4	1			X							
	GP 2.8	1						X				
	GP 2.9	1					X					
	GP 2.10	1					X					

several mappings to organizational PAs, notably OPF SP 1.1 and SP 2.2, which deal with process improvement planning. Taken together, we see how ISO 15288 enhances CMMI-based process improvement by providing a wide range of help when implementing a variety of quality management practices.

The project management PAs are shown in Tables 5.47A and 5.47B and Table 5.48.

What are these maps telling us? Since these PAs are sparsely mapped across so many ISO sections, the resulting matrix becomes quite large. As expected, PP maps very well to section 6.3.1 "Project Planning Process" and PMC maps well to section 6.3.2 "Project Assessment and Control Process." ISO 15288 section 6.2.3 "Project Portfolio Management Process" addresses the collection of projects addressing the strategic objectives of the organization. This process has no direct equivalent in CMMI and is weakly mapped to several PP SPs and the PMC and IPM GPs (mostly GP 2.10). The ISO requirement for prioritizing business opportunities in

Table 5.47A Mapping PP to ISO 15288

	ISO 15288	6.2.2.3 a2	6.2.3.3			6.2.4.3 c3	6.2.5.3 b2	6.3.1.3													6.3.2.3 b5	6.3.4.3 c1	6.3.4.3 e3	6.3.6.3					6.4.1.3 a1
			a2	a5	a6			a1	a2	a3	a4	b1	b2	b3	b4	b5	b6	c1	c2	d2				a1	a2	a3	a4	a5	
CMMI	Total	1	1	1	2	1	1	2	1	1	1	3	2	2	1	1	1	2	1	1	1	1	1	1	1	1	1	1	1
PP SP 1.1	3								X		X										X								
PP SP 1.2	1											X																	
PP SP 1.3	2									X			X																
PP SP 1.4	2											X		X															
PP SP 2.1	5							X				X	X	X									X						
PP SP 2.2	1																					X							
PP SP 2.3	5																							X	X	X	X	X	
PP SP 2.4	4	X													X	X	X												
PP SP 2.5	1					X																							
PP SP 2.6	3			X	X																								X
PP SP 2.7	4		X		X													X	X										
PP SP 3.1	1						X																						
PP SP 3.2	1							X																					
PP SP 3.3	2																	X		X									

Table 5.47B Mapping PMC to ISO 15288

		ISO 15288	*6.1.2.3*		*6.2.4.3 c1*	*6.3.2.3*										*6.3.4.3 e1*
			d1	*d2*		*a1*	*a4*	*a5*	*a6*	*a8*	*a9*	*b2*	*b4*	*b7*	*b8*	
CMMI		Total	10	10	1	7	2	3	2	10	2	1	1	1	1	1
PMC	SP 1.1	7	X	X	X	X	X	X		X						
PMC	SP 1.2	5	X	X		X	X			X						
PMC	SP 1.3	5	X	X		X				X						X
PMC	SP 1.4	4	X	X		X				X						
PMC	SP 1.5	4	X	X		X				X						
PMC	SP 1.6	7	X	X		X		X	X	X	X					
PMC	SP 1.7	7	X	X		X		X	X	X					X	
PMC	SP 2.1	3	X	X						X						
PMC	SP 2.2	6	X	X						X		X	X	X		
PMC	SP 2.3	4	X	X						X	X					

that section is not mapped at all. It may be advantageous for a CMMI-rated organization to explore the portfolio management concept to augment its organizational project management processes.

Now let's take a look at the mapping of IPM and PP to ISO 15288.

What is this map telling us? The mapping of most PP SPs overlaps with IPM, as shown in Table 5.48. The only exception is in the 6.3.1.3 d1 requirement, which deals with obtaining authorization for the project and does not map to any SP. IPM SP 3.4, which addresses establishing the structure of authorities and responsibilities in an integrated team environment, is the only IPPD addition that is mapped. It is supported with the mapping of PP SP 2.4 and GP 2.4 in all PAs. In general, IPM, as an advanced PA, contributes very little to the mapping of the PP PA. ISO 15288, and even with its extensive notes, does not provide additional insights into the project planning process.

With organization-level PAs we have an interesting situation. Let's take Organizational Training and its mapping to ISO 15288 section 6.2.4 "Human Resource Management Processes," shown in Table 5.49. To provide a more complete picture, we show the OT PA supported by mappings to other PAs.

What is this map telling us? In Table 5.49 we see that only the section 6.2.4.3 c4 requirement, which deals with motivation of personnel, is not mapped. Going in the other direction, from CMMI to ISO, we see that OT SP 1.2, which is concerned with determining which training needs are the responsibility of the organization versus projects, and SP 2.3, which deals with training effectiveness, are not mapped to ISO 15288. Adding a process for motivating personnel as required by

Table 5.48 Mapping IPM and PP to ISO 15288 Section 6.3.1

		ISO 15288	*6.3.1.3*													
			a1	*a2*	*a3*	*a4*	*b1*	*b2*	*b3*	*b4*	*b5*	*b6*	*c1*	*c2*	*d2*	*d3*
CMMI		Total	3	1	2	1	5	4	3	2	2	2	3	2	1	1
IPM	SP 1.1	3	X		X			X								
IPM	SP 1.2	1							X							
IPM	SP 1.3	2									X	X				
IPM	SP 1.4	4					X	X					X	X		
IPM	SP 1.5	1														X
IPM	SP 1.6	0														
IPM	SP 2.1	0														
IPM	SP 2.2	1					X									
IPM	SP 2.3	0														
IPM	SP 3.1	0														
IPM	SP 3.2	0														
IPM	SP 3.3	0														
IPM	SP 3.4	1								X						
IPM	SP 3.5	0														
PP	SP 1.1	2		X		X										
PP	SP 1.2	1					X									
PP	SP 1.3	2			X			X								
PP	SP 1.4	2					X		X							
PP	SP 2.1	4	X				X	X	X							
PP	SP 2.2	0														
PP	SP 2.3	0														
PP	SP 2.4	3								X	X	X				
PP	SP 2.5	0														
PP	SP 2.6	0														
PP	SP 2.7	2											X	X		
PP	SP 3.1	0														
PP	SP 3.2	1	X													
PP	SP 3.3	2											X		X	

Table 5.49 Mapping OT to ISO 15288

		ISO 15288	*6.2.4.3*												
			a1	*a2*	*b1*	*b2*	*b3*	*b4*	*c1*	*c2*	*c3*	*c5*	*d1*	*d2*	*d3*
CMMI		Total	1	1	1	1	1	1	1	3	1	3	1	1	1
IPM	SP 2.2	1										X			
IPM	SP 2.3	1										X			
IPM	SP 3.5	1										X			
PP	SP 2.5	1									X				
PMC	SP 1.1	1							X						
OT	SP 1.1	3	X	X						X					
OT	SP 1.2	0													
OT	SP 1.3	4			X					X				X	X
OT	SP 1.4	2				X							X		
OT	SP 2.1	2					X			X					
OT	SP 2.2	1						X							
OT	SP 2.3	0													

Section 6.2.4.3c4, which did not map to CMMI, will certainly enhance a CMMI-based training process.

Other organization level PAs are well represented in ISO 15288, as shown in Table 5.50.

What is this map telling us? Table 5.50 indicates a lot of synergy. CMMI provides a robust training infrastructure, whereas ISO 15288 provides processes for managing personal skills, career development, and motivation, as well as selecting an appropriate knowledge management strategy. From the table one can see that the IPPD addition in the OPD PA (SG 2) and OPD SP 1.4 Establish the Organization's Measurement Repository and OPD SP 1.5 Establish the Organization's Process Asset Library were not mapped and could provide a process foundation for the ISO 15288 processes. In the case of the OPF, several practices did not map, indicating that CMMI has stronger process requirements for deployment of organizational process assets and standard processes, which again could help process deployment for an organization that plans ISO 15288 implementation.

Table 5.50 Mapping OPF and OPD to ISO 15288

		ISO 15288	*6.2.1.3*										*6.2.2.3*		*6.2.3.3*		*6.2.5.3*				
			a1	*a2*	*a3*	*a4*	*a5*	*b1*	*b2*	*b3*	*c1*	*c2*	*a1*	*a2*	*a4*	*a5*	*a1*	*a2*	*a3*	*b2*	*b3*
CMMI		Total	2	3	1	1	1	1	2	1	1	1	2	2	1	1	3	1	1	2	2
OPD	SP 1.1	3		X		X											X				
OPD	SP 1.2	2		X			X														
OPD	SP 1.3	1		X																	
OPD	SP 1.4	0																			
OPD	SP 1.5	0																			
OPD	SP 1.6	3											X	X		X					
OPD	SP 2.1	0																			
OPD	SP 2.2	0																			
OPD	SP 2.3	0																			
OPF	SP 1.1	1																X			
OPF	SP 1.2	1							X												
OPF	SP 1.3	1								X											
OPF	SP 2.1	1									X										
OPF	SP 2.2	1																			X
OPF	SP 3.1	0																			
OPF	SP 3.2	0																			
OPF	SP 3.3	1							X												
OPF	SP 3.4	1										X									
	GP 2.1	2	X														X				
	GP 2.2	2	X														X				
	GP 2.3	3											X	X	X						
	GP 2.4	2			X														X		
	GP 2.5	0																			
	GP 2.6	0																			
	GP 2.7	0																			
	GP 2.8	2						X													X
	GP 2.9	1																		X	
	GP 2.10	1																		X	
	GP 3.1	0																			
	GP 3.2	0																			

Lastly, let's address the high-maturity process areas. The high maturity PAs—OPP, QPM, CAR, and OID—have very sparse maps as shown in Table 5.51.

What is this map telling us? The sparse mapping is due to the ISO 15288 reliance on ISO 15504 for determining process implementation, as noted in Annex B

Table 5.51 Mapping High Maturity PAs to ISO 15288

		ISO 15288	*6.2.1.3 b1*	*6.2.1.3 b3*	*6.2.1.3 c2*	*6.2.5.3 a1*	*6.2.5.3 a2*	*6.2.5.3 b3*	*6.3.1.3 a1*	*6.3.2.3 a7*	*6.3.2.3 b3*
CMMI		Total	1	2	2	0	2	2	1	1	3
OPP	SP 1.1	0									
OPP	SP 1.2	0									
OPP	SP 1.3	1					X				
OPP	SP 1.4	1	X								
OPP	SP 1.5	0									
QPM	SP 1.1	2					X		X		
QPM	SP 1.2	0									
QPM	SP 1.3	0									
QPM	SP 1.4	0									
QPM	SP 2.1	0									
QPM	SP 2.2	0									
QPM	SP 2.3	2						X		X	
QPM	SP 2.4	0									
CAR	SP 1.1	1									X
CAR	SP 1.2	1									X
CAR	SP 2.1	1									X
CAR	SP 2.2	0									
CAR	SP 2.3	0									
OID	SP 1.1	0									
OID	SP 1.2	1		X							
OID	SP 1.3	2		X	X						
OID	SP 1.4	1			X						
OID	SP 2.1	0									
OID	SP 2.2	0									
OID	SP 2.3	1						X			

of the standard. Using CMMI will enable organizations to implement high-maturity concepts as applied to the processes required by ISO 15288.

ISO 12207:2008 to CMMI Maps

As indicated in chapter 4, ISO 12207 is harmonized with ISO 15288, leading to a very strong correspondence between the standards. In general, the mapping relationships follow the cross-references shown in Annex D of ISO 12207. However, from the point of view of mapping to CMMI, there is a much stronger relationship between the standards than the Annex indicates. The differences between the maps are rather small and can be easily pointed out.

The style of this standard follows ISO 9001 in that some shall statements point to whole sections whereas others point to individual statements.

By comparing the 1995 edition to the 2008 revision, one can recognize where the changes occurred and where the statements stayed the same. Therefore, some existing maps, such as maps developed by Griffith University in Australia, can be reused. However, we do caution the users that considering those existing maps must be done very carefully by fully understanding the changes introduced in the standard as well as each map and the underlying assumptions.

The reader will notice that even though ISO 12207 and CMMI cover the similar domains, they are quite different frameworks in their style and coverage. The complete ISO 12207–CMMI map is given in Appendix F.

What can we learn from the ISO 12207 to CMMI mapping? Table 5.52 shows how the similarity between ISO 12207 and ISO 15288 is reflected in their mapping to CMMI. When the ISO 15288 and ISO 12207 requirements are similar, similar maps will be generated. Maps for those sections will not be repeated here. There are, however, unique requirements and requirements that are not addressed by CMMI.

The synergy between CMMI and ISO 12207 is manifested in the details of the ISO standard. For example, the ISO sections dealing with operation, maintenance, domain engineering, and reuse provide an excellent source for addressing topics that are typically not part of a CMMI-based OSSP.

Some of the SPs that did not map are shown in Table 5.53. Those SPs that were discussed in the ISO 15288 section are marked with an asterisk (*).

Table 5.52 High-Level Comparison of ISO 12207, ISO 15288, and CMMI

ISO 12207 Section	*Similarity to ISO 15288*	*CMMI PAs*
6.2.2	Nearly identical	IPM, OPD, PP
6.2.3	Nearly identical	IPM, OPD, PMC, PP, PPQA, RSKM
6.2.4	Nearly identical	IPM, OT
6.2.5	Same	IPM, MA, OID, OPD, OPF,PP, PPQA, QPM
6.3.4	Same	OPF, PMC, PP, RSKM
6.3.5	Same	CM
6.3.6	Same	PMC, PP
6.3.7	Same	MA
6.4.1	Same	PP, RD, REQM

Table 5.53 CMMI Practices That Do Not Map to ISO 12207

PA	SP	Title
IPM	1.2	Use Organizational Process Assets for Planning Project Activities
IPM	1.6*	Contribute to the Organizational Process Assets
IPM	2.1*	Manage Stakeholder Involvement
IPM	2.2	Manage Dependencies
IPM	2.3	Resolve Coordination Issue
OPD	1.3	Establish Tailoring Criteria and Guidelines
OPF	3.1*	Deploy Organizational Process Assets
OPF	3.2*	Deploy Standard Processes
OT	1.2*	Determine Which Training Needs Are the Responsibility of the Organization
PI	1.2	Establish Product Integration Environment
PI	2.1*	Review Interface Descriptions for Completeness
PI	2.2	Manage Interfaces
PI	3.1	Confirm Readiness of Product Components for Integration
REQM	1.1*	Obtain Understanding of Requirements
RSKM	1.3*	Establish a Risk Management Strategy
SAM	1.1	Determine Acquisition Type

* Discussed in the ISO 15288 section.

What is the impact of some of these unmapped topics? It is interesting to note that IPM SG 2 in its entirety is not mapped to ISO 12207, which means that stakeholder coordination will have to be addressed. IPM SP 1.2 and SP 1.6 do not map, which implies that it will be important for an ISO 12207-compliant organization to require that projects contribute artifacts to the repository so that they may be used for planning and estimating purposes. Similar to ISO 15288, Annex D of the standard refers to tailoring of the standard rather than to tailoring of the standard processes. Process tailoring is an important feature in developing and implementing organizational standard processes.

The mapping of ISO 12207 requirements to GPs is rather sparse. Specifically, GP 2.1, GP 2.4, GP 2.5, GP 2.6, and GP 3.1 are seldom mapped. For ISO 12207-based processes, this would have a very strong negative effect on institutionalization.

Going from ISO 12207 to CMMI, the following sections did not map:

- 6.4.9 Software Operation Process
- 6.4.10 Software Maintenance Process
- 6.4.11 Software Disposal Process
- 7.3 Software Reuse Processes (7.3.1 Domain Engineering Process, 7.3.2 Reuse Asset Management Process, 7.3.3 Reuse Program Management Process)

The Operation, Maintenance, and Disposal sections deal with postdeployment aspects of software development and play an important role in completing the software life cycle. Software reuse is briefly mentioned in CMMI but is not explicitly addressed. In each of these cases, ISO 12207 can provide guidelines to CMMI-based processes for addressing those important life cycle aspects.

Similar to the approach used in ISO 15288, each life cycle process discussed is structured to show the process name (section 6.x.y), purpose (6.x.y.1), outcomes (6.x.y.2), and activities and tasks (6.x.y.3). Individual tasks are numbered 6.x.y.3.1, .2, .3, …, and so forth. Since our mapping is performed at the activity/task shall level, only sections 6.x.y.3, and their individual requirements will be mapped. As was the case with ISO 15288, there are several cases where the ISO 12207 standard references requirements in other sections. For example, section 7.2.3.3.1.4 "Execute quality assurance activities" references section 7.2.8 "Software Problem Resolution Process." Those references are included in the mapping tables for the reader's convenience, but no additional mapping information is provided.

In several sections, ISO 12207 indicates that users may implement the corresponding sections of ISO 15288. While we have mapped all requirements, users of the maps, especially those who are implementing CMMI-based process improvement, may take advantage of the standard's compatibility and, using maps as guidance, consider the ISO statements that provide the best additional material.

Similar to ISO 15288, ISO 12207 is a life cycle management standard and, as such, its scope is much closer to CMMI than it is to ISO 9001 or ISO 20000. Therefore, some organizations use ISO 12207 as a supplement or source of best practices without concern for compliance or conformance with the standard. In the following discussion, we will point to some areas that are particularly suitable for those purposes. We will also indicate those areas where an ISO-compliant organization will have to rely on CMMI as its source of institutionalization.

As indicated in Table 5.52, the ISO 12207 sections that have identical maps to CMMI as those in ISO 15288 are:

- 6.2.5 Quality Management Process
- 6.3.4 Risk Management Process
- 6.3.5 Configuration Management Process
- 6.3.6 Information Management Process
- 6.3.7 Measurement Process
- 6.4.1 Stakeholder Requirements Definition Process

Amplifications to sections 6.2.5 and 6.3.5 are provided in sections 7.2.3 "Software Quality Assurance Process" and 7.2.2 "Software Configuration Management Process," respectively.

An example of CMMI MA PA mapping to both ISO 12207 and 15288 is shown in Table 5.54.

Table 5.54 Mapping of MA to ISO 15288 and 12207

		ISO 15288: 6.3.7.3	*a1*	*a2*	*a3*	*a4*	*a5*	*a6*	*a7*	*b1*	*b2*	*b3*	*b4*	*c1*	*c2*
		ISO 12207: 6.3.7.3	6.3.7.3.1.1	6.3.7.3.1.2	6.3.7.3.1.3	6.3.7.3.1.4	6.3.7.3.1.5	6.3.7.3.1.6	6.3.7.3.1.7	6.3.7.3.2.1	6.3.7.3.2.2	6.3.7.3.2.3	6.3.7.3.2.4	6.3.7.3.3.1	6.3.7.3.3.2
CMMI		Total	1	1	1	1	1	1	1	1	2	1	1	1	2
MA	SP 1.1	2	X	X											
MA	SP 1.2	1			X										
MA	SP 1.3	1				X									
MA	SP 1.4	1					X								
MA	SP 2.1	1									X				
MA	SP 2.2	1										X			
MA	SP 2.3	1									X				
MA	SP 2.4	1											X		
MA	GP 2.2	1								X					
MA	GP 2.3	2						X	X						
MA	GP 2.9	2												X	X
MA	GP 3.2	1													X

We will first consider the Engineering PAs, which exhibit quite a strong relationship to ISO 12207. The RD and REQM PAs are mapped to section 6.4.1 and the maps are identical to those for ISO 15288 and will not be repeated here. The mapping of ISO 12207 to CMMI covers the full lifecycle from supply, implementation, design, construction, and integration, providing a process thread that is not as clearly visible or articulated in CMMI.

The TS PA mapping is shown in Table 5.55.

What is this map telling us? As we can see, all TS SPs are mapped to ISO requirements that interact iteratively covering the Acquisition (6.1.1), Supply (6.1.2), System Architectural Design (6.4.3), Software Architectural Design (7.1.3), Software Detailed Design (7.1.4), Software Construction (7.1.5), Software Integration (7.1.6), Software Qualification (7.1.7), Software Documentation Management (7.2.1, updating documentation), and Software Verification (7.2.4, verifying documentation) processes. The ISO 12207 section 7.1.3 and 7.1.4 requirements are more detailed than their CMMI counterparts. The mapping covers all aspects of design and implementation, as well as reviews, testing, and documentation.

Now let's consider PI, as shown in Table 5.56.

What is this map telling us? As we can see from Table 5.56, not all PI SPs are mapped and those SPs that do map are spread over a large number of ISO 12207

Table 5.55 Mapping TS to ISO 12207

		ISO12207	*6.1.1.3.1.6*	*6.1.1.3.1.7*	*6.1.2.3.4.4*	*6.4.3.3.1.1*	*6.4.3.3.2.1*	*7.1.3.3.1.1*	*7.1.3.3.1.2*	*7.1.3.3.1.3*	*7.1.3.3.1.4*	*7.1.3.3.1.6*	*7.1.4.3.1.1*	*7.1.4.3.1.2*	*7.1.4.3.1.3*	*7.1.4.3.1.4*	*7.1.4.3.1.7*	*7.1.5.3.1.1*	*7.1.5.3.1.2*	*7.1.5.3.1.3*	*7.1.5.3.1.5*	*7.1.6.3.1.3*	*7.1.7.3.1.2*	*7.2.1.3.2.2*	*7.2.4.3.2.2*	*7.2.4.3.2.3*	*7.2.4.3.2.5*
CMMI		Total	2	1	2	3	2	1	1	1	1	1	1	1	1	1	1	1	1	1	1	1	1	2	1	1	2
TS	SP 1.1	5	X	X	X	X	X																				
TS	SP 1.2	2				X	X																				
TS	SP 2.1	9				X		X	X	X		X	X		X		X								X		
TS	SP 2.2	2																						X			X
TS	SP 2.3	1												X													
TS	SP 2.4	2	X		X																						
TS	SP 3.1	4																X	X		X					X	
TS	SP 3.2	7									X					X				X		X	X	X			X

Table 5.56 Mapping PI to ISO 12207

		ISO 12207	6.1.2.3.5.1	6.1.2.3.6.2	6.4.5.3.1.1	6.4.6.3.1.4	7.1.3.3.1.5	7.1.4.3.1.6	7.1.5.3.1.4	7.1.6.3.1.1	7.1.6.3.1.2	7.1.6.3.1.4	7.1.6.3.1.5	7.2.4.3.2.4	7.2.4.3.2.5
CMMI		Total	1	1	2	1	1	1	2	2	2	1	1	1	1
PI	SP 1.1	2							X	X					
PI	SP 1.2	0													
PI	SP 1.3	5					X	X	X	X		X			
PI	SP 2.1	0													
PI	SP 2.2	0													
PI	SP 3.1	0													
PI	SP 3.2	2			X						X				
PI	SP 3.3	4			X						X		X	X	
PI	SP 3.4	3	X	X		X									

requirements. The PI practices that do not map to ISO 12207—SP 1.2 Establish the Product Integration Environment, SP 2.1 Review Interface Descriptions for Completeness, SP 2.2 Manage Interfaces, and SP 3.1 Confirm Readiness of Product Components for Integration—will have to be specifically addressed in an ISO-compliant organization. These practices, particularly SP 2.1 and SP 2.2, bring additional rigor to product integration.

The maps for Validation (VAL) and Verification (VER) are shown in Table 5.57 and Table 5.58.

What are these maps telling us? Here we see that the VAL PA is mapped completely to ISO 12207 section 7.2.5 and VER is mapped across many ISO 12207 sections. This is because there are many instances where the verification process is invoked in ISO 12207, for example, in the Technical Processes (section 6.4), Software Life Cycle Processes (section 7.1), and, of course, in Software Verification Process (section 7.2.4) essentially following the life cycle. A closer inspection of the section 7.2.4 mapping (see Appendix F) indicates that all requirements in section 7.2.4.3.2 are mapped to other Engineering PAs rather than to VER.

It is interesting to note that ISO 12207 section 7.1.7, Software Qualification Testing, has two requirements for conducting qualification testing and evaluating design, code, and documentation, which map to VER SP 3.1 and 3.2. The requirements specifically address testing as opposed to other verification techniques such as analysis or reviews. For CMMI-based process improvement, this provides a wealth of practices that can supplement the CMMI VER and VAL PAs.

Next, we'll discuss the CM and PPQA PAs as shown in Table 5.59 and Table 5.60.

Table 5.57 Mapping VAL to ISO 12207

CMMI		*ISO 12207* Total	*7.2.5.3.1.1*	*7.2.5.3.1.2*	*7.2.5.3.1.4*	*7.2.5.3.1.5*	*7.2.5.3.2.1*	*7.2.5.3.2.2*	*7.2.5.3.2.3*	*7.2.5.3.2.4*	*7.2.5.3.2.5*
			1	1	3	2	3	1	1	1	1
VAL	SP 1.1	2	X				X				
VAL	SP 1.2	1					X				
VAL	SP 1.3	1					X				
VAL	SP 2.1	3				X			X		X
VAL	SP 2.2	2				X				X	
VAL	GP 2.1	0									
VAL	GP 2.2	2		X	X						
VAL	GP 2.3	1			X						
VAL	GP 2.4	1			X						
VAL	GP 2.5	0									
VAL	GP 2.6	0									
VAL	GP 2.7	0									
VAL	GP 2.8	0									
VAL	GP 2.9	1						X			
VAL	GP 2.10	0									
VAL	GP 3.1	0									
VAL	GP 3.2	0									

Table 5.58 Mapping VER to ISO 12207

CMMI		*ISO 12207* Total	*6.4.2.3.2.1*	*6.4.3.3.2.1*	*6.4.5.3.1.1*	*6.4.5.3.2.1*	*6.4.6.3.1.1*	*6.4.6.3.1.2*	*7.1.2.3.1.2*	*7.1.3.3.1.5*	*7.1.3.3.1.6*	*7.1.4.3.1.5*	*7.1.4.3.1.6*	*7.1.4.3.1.7*	*7.1.5.3.1.4*	*7.1.5.3.1.5*	*7.1.6.3.1.5*	*7.1.7.3.1.1*	*7.1.7.3.1.3*	*7.2.1.3.2.3*	*7.2.3.3.2.3*	*7.2.4.3.1.1*	*7.2.4.3.1.4*	*7.2.4.3.1.6*	*7.2.6.3.3.1*
			3	3	1	3	1	1	3	1	3	2	1	2	1	3	8	1	1	3	2	1	1	5	5
VER	SP 1.1	4										X					X					X	X		
VER	SP 1.2	1															X								
VER	SP 1.3	6				X				X		X	X		X		X								
VER	SP 2.1	9	X	X					X		X					X	X			X				X	X
VER	SP 2.2	9	X	X					X		X					X	X			X				X	X
VER	SP 2.3	9	X	X					X		X					X	X			X				X	X
VER	SP 3.1	9			X	X	X							X			X	X			X			X	X
VER	SP 3.2	8				X		X						X			X		X		X			X	X

Table 5.59 Mapping CM to ISO 12207

CMMI	*ISO 12207*	*6.3.5.3.1.1*	*6.3.5.3.1.2*	*6.3.5.3.2.1*	*6.3.5.3.2.2*	*7.2.1.3.4.1*	*7.2.2.3.2.1*	*7.2.2.3.3.1*	*7.2.2.3.4.1*	*7.2.2.3.5.1*	*7.2.2.3.6.1*
CMMI	Total	1	1	3	4	7	1	2	1	1	1
CM SP 1.1	3		X			X	X				
CM SP 1.2	2	X				X					
CM SP 1.3	3			X		X					X
CM SP 2.1	4			X	X	X		X			
CM SP 2.2	3				X	X		X			
CM SP 3.1	4			X	X	X			X		
CM SP 3.2	3				X	X				X	

Table 5.60 Mapping PPQA to ISO 12207

CMMI	*ISO 12207*	*6.2.5.3.2.1*	*6.2.5.3.2.2*	*6.3.2.3.3.1*	*7.2.1.3.2.3*	*7.2.3.3.1.4*	*7.2.3.3.1.5*	*7.2.3.3.1.6*	*7.2.3.3.2.1*	*7.2.3.3.2.2*	*7.2.3.3.2.3*	*7.2.3.3.3.1*	*7.2.3.3.3.2*	*7.2.3.3.3.5*	*7.2.3.3.3.6*	*7.2.7.3.1.2*	*7.2.7.3.1.5*	*7.2.7.3.1.6*	*7.2.7.3.1.7*	*7.2.7.3.2.1*
CMMI	Total	1	1	1	1	4	1	2	1	1	1	1	1	2	2	2	1	1	1	2
PPQA SP 1.1	8					X		X				X	X	X	X	X				X
PPQA SP 1.2	11			X	X	X		X	X	X	X			X	X	X				X
PPQA SP 2.1	5	X				X											X	X	X	
PPQA SP 2.2	3		X			X	X													

What are these maps telling us? In Table 5.59 we see that CM is mapped to ISO section 6.3.5 "Configuration Management Process" and to section 7.2.2 "Software Configuration Management Process." ISO 12207 refers to software configuration management as a "specialization" of the process in section 6.3.5. Those two ISO sections together provide very complete guidelines for CM in CMMI-based process improvement.

In the case of PPQA we have a diverse mapping that can support CMMI-based process improvement very well. Here we see that PPQA SPs map to Quality Management Process (section 6.2.5), Software Quality Assurance (section 7.2.3), and Software Audit Process (section 7.2.7) with an additional cross-reference to the Software Review Process (section 7.2.8). All of these ISO requirements map to just four PPQA SPs, thus providing an excellent source of best practices for CMMI-based process improvement.

Let's now consider the Project Management PAs, namely, PP and PMC by inspecting Table 5.61 and Table 5.62.

Table 5.61 Mapping PP to ISO 12207

		ISO 12207	*6.1.2.3.4.2*	*6.1.2.3.4.5*	*6.1.2.3.4.10*	*6.1.2.3.4.11*	*6.1.2.3.4.12*	*6.2.3.3.1.2*	*6.2.3.3.1.3*	*6.2.3.3.1.5*	*6.2.3.3.1.6*	*6.2.4.3.1.1*	*6.2.4.3.1.2*	*6.2.4.3.2.1*	*6.2.4.3.3.8*	*6.2.5.3.1.5*	*6.3.1.3.1.1*	*6.3.1.3.1.2*	*6.3.1.3.1.3*	*6.3.1.3.2.1*	*6.3.1.3.3.2*	*6.3.4.3.3.1*	*6.3.4.3.5.3*	*6.3.6.3.1.1*	*6.3.6.3.1.2*	*6.3.6.3.1.3*	*6.3.6.3.1.4*	*6.3.6.3.1.5*	*6.4.1.3.1.1*	*7.1.1.3.1.1*	*7.2.6.3.1.2*
CMMI		Total	1	1	1	1	1	1	1	1	2	1	1	1	1	1	1	1	1	14	1	1	1	1	1	1	1	1	1	1	1
PP	SP 1.1	3							X								X			X											
PP	SP 1.2	1																		X											
PP	SP 1.3	3	X																	X										X	
PP	SP 1.4	1																		X											
PP	SP 2.1	2																		X			X								
PP	SP 2.2	2																		X		X									
PP	SP 2.3	6																		X				X	X	X	X	X			
PP	SP 2.4	2																		X											X
PP	SP 2.5	5										X	X	X	X					X											
PP	SP 2.6	7			X	X	X			X	X									X									X		
PP	SP 2.7	4		X				X			X									X											
PP	SP 3.1	3														X		X		X											
PP	SP 3.2	3																	X	X	X										
PP	SP 3.3	1																		X											

Table 5.62 Mapping PMC to ISO 12207

CMMI	ISO 12207	6.1.2.3.4.6	6.1.2.3.4.8	6.1.2.3.4.10	6.1.2.3.4.11	6.1.2.3.4.12	6.3.1.3.3.1	6.3.2.3.1.1	6.3.2.3.2.1	6.3.2.3.2.2	6.3.4.3.5.1	6.3.6.3.2.1	6.3.6.3.2.2	6.3.6.3.2.3	6.3.6.3.2.4	6.3.6.3.2.5	6.3.6.3.2.6	7.1.7.3.1.5	7.2.6.3.1.1	7.2.6.3.1.4	7.2.6.3.1.5	7.2.6.3.1.6	7.2.6.3.2.1	7.2.6.3.3.1	7.2.8.3.2.1
	Total	10	10	1	1	2	7	5	3	2	1	1	1	1	1	1	1	3	1	1	1	1	3	1	3
PMC SP 1.1	4	X	X				X	X																	
PMC SP 1.2	4	X	X				X	X																	
PMC SP 1.3	6	X	X				X	X			X												X		
PMC SP 1.4	10	X	X				X	X				X	X	X	X	X	X								
PMC SP 1.5	7	X	X	X	X	X	X	X																	
PMC SP 1.6	5	X	X				X			X													X		
PMC SP 1.7	10	X	X			X	X			X									X		X	X	X	X	
PMC SP 2.1	6	X	X						X									X		X					X
PMC SP 2.2	5	X	X						X									X							X
PMC SP 2.3	5	X	X						X									X							X

What are these maps telling us? Both PP and PMC are fully mapped to ISO 12207. These maps are nearly identical to the ISO 15288 mapping described earlier in this chapter. The difference is in the addition of the mapping to the Software Life Cycle processes in ISO 12207 section 7. The PP mapping covers many sections of the ISO standard, including the Supply (section 6.1.2), Project Portfolio Management (section 6.2.3), Human Resource Management (section 6.2.4), Project Planning (section 6.3.1), Risk Management (section 6.3.4), and Information Management (section 6.3.6) processes as well as several Software Life Cycle processes related to selection of life cycle and obtaining agreement of resources for reviews.

Similarly, the PMC mapping covers the Supply, Project Assessment and Control (section 6.3.2), Risk Management (section 6.3.4), Information Management (section 6.3.6), and Software Review (section 7.2.6) processes, as well as processes for updating and preparing software product for integration, testing, and acceptance. As expected, those processes that were introduced in planning are monitored and controlled.

Finally, the relationships of the organizational-level PAs to ISO 12207 are shown in Table 5.63.

Table 5.63 OPD and OPF Mapping to ISO 12207

		ISO 12207	*6.1.2.3.4.3*	*6.2.1.3.1.1*	*6.2.1.3.2.1*	*6.2.1.3.3.1*	*6.2.1.3.3.2*	*6.2.1.3.3.3*	*6.2.2.3.1.1*	*6.2.2.3.2.2*	*6.2.2.3.3.1*	*6.2.3.3.1.5*	*6.2.4.3.3.6*	*6.2.4.3.3.7*	*6.2.4.3.4.3*	*6.2.4.3.4.5*	*6.2.5.3.1.1*	*6.2.5.3.1.2*	*6.2.5.3.1.6*	*7.2.1.3.2.1*	*7.2.3.3.4.1*
CMMI		Total	1	3	1	3	2	2	3	3	1	1	1	1	1	1	1	1	1	1	1
OPD	SP 1.1	4	X	X													X			X	
OPD	SP 1.2	1		X																	
OPD	SP 1.3	0																			
OPD	SP 1.4	3						X	X	X											
OPD	SP 1.5	4							X	X					X	X					
OPD	SP 1.6	4							X	X	X	X									
OPD	SP 2.1	1												X							
OPD	SP 2.2	1											X								
OPD	SP 2.3	0																			
OPF	SP 1.1	1																X			
OPF	SP 1.2	2			X																X
OPF	SP 1.3	1				X															
OPF	SP 2.1	1				X															
OPF	SP 2.2	2				X													X		
OPF	SP 3.1	0																			
OPF	SP 3.2	0																			
OPF	SP 3.3	1					X														
OPF	SP 3.4	3		X			X	X													

What is this map telling us? The maps shown here are quite similar to the ISO 15288 maps although the wording of the requirements differs. For example, ISO 12207 provides more elaborate requirements for the definition and implementation of infrastructure in section 6.2.2 that correspond quite well to the OPD PA.

However, several SPs in the OPD and OPF PAs show no relationships to ISO 12207 requirements. One significant practice that did not map is OPD SP 1.3 Establish Tailoring Guidelines and Criteria, which has a major impact on establishing and implementing a project's defined processes and the mapping of IPM SP 1.1 and GP 3.1 in all PAs. Maps from IPM SP 1.1 only address selecting a life cycle model rather than establishing a defined process. Thus, CMMI is more specific and provides more guidance for implementing the process infrastructure. OPF SPs for deploying process assets and deploying standard processes have no corresponding ISO 12207 requirements. Here too, CMMI is more detailed and more useful in implementing process improvement guidelines.

Let's now consider training, as shown in Table 5.64.

What is this map telling us? The OT PA is largely satisfied by ISO 12207 section 6.2.4 "Human Resource Management Process," but two OT SPs remain unsatisfied, such as determining which training needs are the responsibility of the organization, aspects of delivering training, and assessing training effectiveness. Going in the other direction, from ISO 12207 to CMMI, many ISO requirements are not supported. The requirements such as skill acquisition and provision (Section 6.2.4.3.3) and knowledge management (Section 6.2.4.3.4) could greatly enhance CMMI-based training capability.

Last, let's address the high-maturity process areas. We already indicated that CAR did not map at all. The other three high maturity PAs—OPP, QPM, and OID—have very sparse maps as shown in Table 5.65.

Table 5.64 OT Mapping to ISO 12207

		ISO 12207	6.2.4.3.1.1	6.2.4.3.1.2	6.2.4.3.2.1	6.2.4.3.2.2	6.2.4.3.2.3	6.2.4.3.3.5	6.2.4.3.3.8
CMMI		Total	1	1	1	1	1	1	1
OT	SP 1.1	3	X	X					X
OT	SP 1.2	0							
OT	SP 1.3	1			X				
OT	SP 1.4	1				X			
OT	SP 2 1	1					X		
OT	SP 2.2	1						X	
OT	SP 2.3	0							

Table 5.65 High Maturity PA Mapping to ISO 12207

CMMI		*ISO 12207* Total	6.2.5.3.1.2 2	6.2.5.3.1.6 2
OID	SP 1.1	0		
OID	SP 1.2	0		
OID	SP 1.3	0		
OID	SP 1.4	0		
OID	SP 2.1	0		
OID	SP 2.2	0		
OID	SP 2.3	1		X
OPP	SP 1.1	0		
OPP	SP 1.2	0		
OPP	SP 1.3	1	X	
OPP	SP 1.4	0		
OPP	SP 1.5	0		
QPM	SP 1.1	1	X	
QPM	SP 1.2	0		
QPM	SP 1.3	0		
QPM	SP 1.4	0		
QPM	SP 2.1	0		
QPM	SP 2.2	0		
QPM	SP 2.3	1		X
QPM	SP 2.4	0		

What is this map telling us? ISO 12207 Annex B, Process Reference Model (PRM) for Assessment Purposes, provides "process attributes" suitable for appraising process implementation. Three attributes discussed in that annex (PA 4.2 process control, PA 5.1 process innovation, and PA 5.2 process optimization) deal with quantitative aspects of process implementation, but there is nothing in the body of the standard that addresses quantitatively managed and optimizing processes. Those attributes are equivalent to CMMI GP 4.2 and GP 5.2, although the ISO standard makes much less detail available. Therefore, an organization that would like to comply with ISO 12207 by using those attributes may find it useful to implement either GG 4 and GG 5, or the OPP, QPM, CAR, and OID PAs.

Summary

In this chapter we have presented the maps relating CMMI v1.2 to each of the selected ISO standards using a CMMI-centric approach. For each shall statement in the ISO standard, the appropriate CMMI SPs or GPs are identified that, when implemented, will satisfy that ISO requirement. To more precisely characterize mappings, we used a confidence factor for each mapping to distinguish between strong and weak relationships. Several PAs have very strong maps, which are localized to a selected set of ISO requirements, and others have very sparse maps where the relationships between PA practices and ISO requirements are scattered over a large number of instances. Thus, some difficulties in developing a consistent process improvement approach may be encountered. We conclude that CMMI can guide an organization in establishing and maintaining a strong process infrastructure while addressing the specific requirements outlined in the ISO standards. Additional ISO-required processes may have to be developed when there is no corresponding process in CMMI. We encourage users of the maps, especially those who are implementing CMMI-based process improvement, to take advantage of framework compatibility and then, using maps as guidance, consider the ISO requirements in the registration process or as best practices in process improvement.

In the next chapter, we will show how the individual maps may be brought together and used for effective process improvement.

Chapter 6

Tying It All Together

After completing the mapping exercise, and after considering our experiences in performing process improvement consulting and appraisals, we realized that a major concept emerging from the maps is that they are much better tools for assessing compliance than for guiding process improvement. Process improvement is more than concentrating on the map for an individual requirement, let alone caring if it's a weak, medium, or strong quality map. It requires a much broader look at the total environment and relevant business goals.

An important point to remember when using maps for process improvement is that organizations do not implement individual practices—they implement processes where those practices may be individual tasks or activities. This means that, in general, even if only one practice is mapped to a requirement, the whole PA, or an equivalent set of processes, will have to be developed, implemented, measured, and improved. In addition, the relationships, and specifically temporal relationships, that are so important in process development and improvement, may be lost in the mapping.

When discussing the process improvement aspects of mapping, we will describe an approach that will accommodate all of the selected ISO standards. We will also attempt to point out differences that occur when an ISO-registered organization attempts to achieve a CMMI capability or maturity level and when a CMMI-rated organization attempts to achieve ISO registration.

We will also develop the relationships between several ISO standards and CMMI and between two widely used ISO standards: ISO 9001 and ISO 20000.

Basic Concepts

Process improvement requires a much deeper understanding of the frameworks than is needed for assessing compliance, although it helps if the appraisers fully understand all the frameworks that are being considered when assessing compliance.

To facilitate our discussion of using multiple frameworks for process improvement, we use a CMMI-centric approach. In chapter 5 we described individual ISO-to-CMMI maps and their process improvement interpretation. In this chapter, using those maps, we will create a process view that includes CMMI and multiple ISO standards. We will first discuss the CMMI PA-to-GP relationships and their effect on the use of multiple frameworks, as we believe that these institutionalization practices represent a cornerstone of successful process improvement. We will also follow the CMMI constellation approach (described in chapter 3), where core PAs are identified in the model foundation and other PAs are added as part of the constellation-specific material.

This approach enables us to explain the interaction of seemingly divergent frameworks and to point out their synergy. This approach also exposes major differences between CMMI and the ISO standards considered in this book.

Using Multiple Framework Relationships for Process Improvement

To explain how to use multiple frameworks for process improvement, we will use the generic process architecture shown in Figure 6.1. This architecture has been selected for illustrative purposes only; it is *not* a recommended architecture. We will provide several examples to show how frameworks interact and show cases

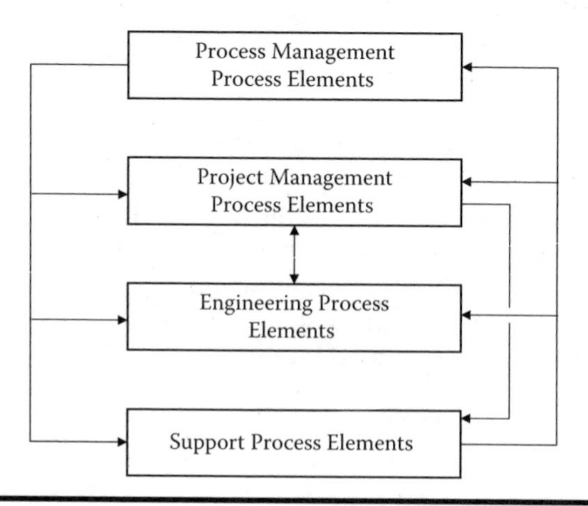

Figure 6.1 Generic high-level process architecture.

where weaknesses in one or more frameworks are supplemented by strengths in other frameworks.

We have selected four major groupings of processes corresponding to the CMMI continuous representation categories and have indicated their high-level interactions. Each process grouping is further decomposed into subprocesses and finally into process elements (PEs). According to CMMI, a PE is defined as a fundamental unit of a process, where each PE covers a closely related set of activities (or tasks), has inputs, and creates output work products by using resources. When dealing with multiple frameworks, the process architect will consider all relevant frameworks when creating PEs and will use the architecture to identify relationships and interfaces among PEs. Because of PE cohesion and coupling considerations, creating an architecture is generally an iterative process, and certain framework requirements may be moved between PEs several times in the course of process development before selecting the best allocation.

Our approach to showing framework interaction is CMMI-centric; we start with CMMI practices to provide the organizing concepts and then bring other frameworks into consideration. Alternatively, an ISO 9001-registered organization may start with the ISO standard and then, using the same approach, add other frameworks when developing process definitions and descriptions. We also capitalize on the CMMI architecture described in chapter 3, specifically using the core CMMI PAs. As we have shown in chapter 5, the engineering processes described in ISO 20000 have rather weak maps to CMMI while, as expected, their project management, support, and process management processes exhibit much stronger maps to the CMMI core PAs. This translates into additional requirements for developing the specific engineering processes needed to satisfy CMMI expectations while still preserving project management and supporting processes.

In chapter 2 we described two well-known approaches to process improvement, namely, PDCA (Plan–Do–Check–Act) and IDEAL. As we review the frameworks and their joint maps to CMMI, we will keep in mind that a process improvement approach will have to be selected and used by process architects and process improvement groups to guide them in creating and implementing those new processes. ISO standards refer to PDCA in their text, whereas CMMI leaves the choice of process improvement approach to the implementers. We feel strongly that a defined approach should be considered for planning, developing, implementing, and deploying processes. Our choice is IDEAL since it not only shows how to plan and implement changes, but also indicates steps for addressing the business need for process improvement and obtaining lessons learned that may become guidance for the next process improvement cycle (Mutafelija and Stromberg 2003).

When process architects decide to use all the frameworks analyzed in this book, they will notice differences in the granularity of the frameworks. ISO 15288 and ISO 12207 are detailed with many requirements and notes providing implementation hints and cross-references to other standards. ISO 9001 and ISO 20000 are at a much higher level, listing fewer but more general requirements. CMMI is somewhere in the middle, providing a balanced set of practices. This granularity plays a major role in mapping among the frameworks and in their value in process improvement and compliance determination. For example, ISO 12207 has several sections that deal with reviews, audits, and different types of testing which, if considered as best practices, will provide guidance when developing the related process elements.

The interaction among multiple frameworks will affect the process architecture, particularly where certain activities are not present in all frameworks. Those considerations will typically be addressed by the organizational business goals guiding process development and improvement. From our experience, we see those frameworks and their maps as requirements for process elements that should be considered in their totality. This approach helps to create compact and cohesive process elements with minimal coupling, which will generate work products and outcomes satisfying multiple frameworks.

In this chapter, we will first show several examples of CMMI mapping to both ISO 9001 and ISO 20000, followed by several examples of CMMI mapping to both ISO 12207 and ISO 15288, and, finally, several mappings of CMMI to all ISO frameworks. In the discussion of CMMI mapping to ISO 9001 and ISO 20000, we will frequently address the relationships between ISO 9001 and ISO 20000.

In this chapter, as in chapter 5, we rely on comparison matrices to convey framework interaction. However, a process architect must also consider the relative strength of each mapping, expressed as a confidence percentage. In certain cases, a mapping may be very weak, covering only a portion of a requirement. In that case, the process architect may explore several options ranging from developing a process that satisfies the requirement to deciding not to implement the requirement at all.

CMMI Mapping to ISO 9001 and ISO 20000

The maps for the PP and PMC project management processes are shown in Table 6.1 and Table 6.2. The maps in this chapter have an additional column, labeled "Total 9001/20000," that provides sum of the individual maps. We use this column as an indicator of the synergy among the maps.

What are these maps telling us? For PP, the two practices (SP 1.2 and SP 1.3) that are not mapped to ISO 20000 are covered by the mapping to ISO 9001. This implies that an organization that has implemented the CMMI PP PA would provide estimates for work products and definition of project life cycles to satisfy both ISO standards. Similarly, an organization that has implemented the PMC PA would monitor relevant planning parameters, commitments, and risks that would satisfy both standards. Similarly, an ISO 20000-registered organization will have processes in place for monitoring project planning parameters, commitments, and risks, which may be, with some interpretation, extended to ISO 9001 to cover those missing tasks.

From the engineering category, we will start with the requirements processes shown in Table 6.3.

Table 6.1 Mapping of PP to ISO 9001 and ISO 20000

			ISO 9001	*4.2.1*	*4.2.3*	*6.2.2*	*6.3*	*6.4*	*7.1*	*7.3.1*	*ISO 20000*	*3.2*	*3.3*	*4.1*	*4.2*	*5*	*6.4*	*6.5*	*6.6*	*7.2*
CMMI		Total 9001/ 20000	Total	1	1	1	1	1	4	14	Total	1	1	6	1	8	5	1	1	1
PP	SP 1.1	4	1							X	3			X		X	X			
PP	SP 1.2	1	1							X	0									
PP	SP 1.3	2	2						X	X	0									
PP	SP 1.4	3	1							X	2					X	X			
PP	SP 2.1	4	1							X	3			X		X	X			
PP	SP 2.2	3	1							X	2				X				X	
PP	SP 2.3	5	4	X	X				X	X	1	X								
PP	SP 2.4	6	3				X	X		X	3			X		X		X		
PP	SP 2.5	4	2			X				X	2		X			X				
PP	SP 2.6	3	1							X	2					X				X
PP	SP 2.7	3	2						X	X	1			X						
PP	SP 3.1	3	2						X	X	1			X						
PP	SP 3.2	3	1							X	2					X	X			
PP	SP 3.3	4	1							X	3			X		X	X			

Table 6.2 Mapping of PMC to ISO 9001 and ISO 20000

		Total 9001/	*ISO 9001*	*4.2.3*	*7.3.4*	*8.2.1*	*8.2.3*	*8.3*	*8.4*	*8.5.2*	*8.5.3*	*ISO 20000*	*3.3*	*4.2*	*4.3*	*6.1*	*6.2*	*6.4*	*6.6*	*7.2*	*7.3*	*7.3*
CMMI		20000	Total	1	3	1	3	3	1	3	3	Total	1	4	10	3	1	1	2	1	2	2
PMC	SP 1.1	4	0									4	X	X	X			X				
PMC	SP 1.2	1	0									1			X							
PMC	SP 1.3	3	0									3		X	X				X			
PMC	SP 1.4	2	1	X								1			X							
PMC	SP 1.5	3	2			X			X			1			X							
PMC	SP 1.6	3	1		X							2		X	X							
PMC	SP 1.7	6	1		X							5		X	X					X	X	X
PMC	SP 2.1	9	5		X		X	X		X	X	4			X	X	X		X			
PMC	SP 2.2	8	4				X	X		X	X	4			X	X					X	X
PMC	SP 2.3	6	4				X	X		X	X	2			X	X						

Table 6.3 Mapping RD and REQM to ISO 9001 and ISO 20000

		Total 9001/	*ISO 9001*	*7.1*	*7.2.1*	*7.2.2*	*7.2.3*	*7.3.1*	*7.3.2*	*7.3.3*	*7.6*	*ISO 20000*	*3.2*	*4.1*	*5*	*6.1*	*6.3*	*6.4*	*6.5*	*6.6*	*7.2*
CMMI		20000	Total	4	6	9	5	0	9	0	0	Total	3	1	4	9	3	1	2	2	3
RD	SP 1.1	5	3		X		X		X			2				X					X
RD	SP 1.2	9	5	X	X	X	X		X			4	X			X	X				X
RD	SP 2.1	8	4	X	X	X			X			4	X		X	X	X				
RD	SP 2.2	1	0									1					X				
RD	SP 2.3	4	3	X	X				X			1		X							
RD	SP 3.1	3	3		X		X		X			0									
RD	SP 3.2	3	3	X	X				X			0									
RD	SP 3.3	4	2			X			X			2			X	X					
RD	SP 3.4	4	2			X			X			2			X	X					
RD	SP 3.5	4	3			X	X		X			1				X					
REQM	SP 1.1	5	2			X	X					3				X			X	X	
REQM	SP 1.2	3	1			X						2	X			X					
REQM	SP 1.3	7	1			X						6			X	X		X	X	X	X
REQM	SP 1.4	0	0									0									
REQM	SP 1.5	1	1			X						0									

What is this map telling us? One CMMI practice, REQM SP 1.4 Maintain Bidirectional Traceability of Requirements, has no mapping to either ISO standard. Organizations using both frameworks would benefit from including the requirements traceability advocated by CMMI. An ISO 9001-registered organization may provide a solid infrastructure for requirements development on service projects, specifically though establishing a concept of operation and a definition of required functionality (ISO 9001 section 7.3.2), and by identifying inconsistencies between requirements and other work products (ISO 9001 section 7.2.2).

Recalling the CMMI constellation concept when examining the TS and PI PAs, we can see that ISO 20000 requirements are so specific to the service domain that neither CMMI nor ISO 9001 will satisfy those requirements. The ISO 20000 processes will have to be specifically addressed.

In the case of the VER and VAL PAs, we have a different situation, as shown in Table 6.4 and Table 6.5.

What are these maps telling us? Here we have a much better correspondence than in the TS and PI PAs. For verification, CMMI and ISO 9001 provide a solid infrastructure, specifically including peer reviews. This would certainly benefit an organization that plans its processes based on ISO 20000. For validation, an organization that is ISO 20000 certified may also benefit from using CMMI or ISO 9001 or both to establish a process for selecting products for validation.

Let's now turn our attention to support processes. The mapping of the CMMI MA PA to ISO 9001 and ISO 20000 is shown in Table 6.6.

What is this map telling us? Here we see solid mappings among all three frameworks. Specifically, ISO 9001 section 5.4.1 and ISO 20000 sections 4.3, 6.5, and 6.6 require organizations to determine measurement objectives linked to their quality goals that will lead to selection, collection, and analysis of measurements.

Table 6.4 Mapping VER to ISO 9001 and 20000

CMMI	Total 9001/ 20000	*ISO 9001* Total	*6.3* 1	*6.4* 1	*7.3.1* 1	*7.3.3* 1	*7.3.5* 8	*7.4.3* 1	*7.5.3* 1	*7.6* 1	*8.1* 5	*8.2.4* 8	*8.3* 1	*8.4* 1	*8.5.2* 1	*8.5.3* 1	*ISO 20000* Total	*5* 3	*6.1* 1	*6.2* 1	*10.1* 1
VER SP 1.1	4	4			X		X				X	X					0				
VER SP 1.2	6	5	X	X			X				X	X					1				X
VER SP 1.3	6	5				X	X			X	X	X					1	X			
VER SP 2.1	2	2					X					X					0				
VER SP 2.2	2	2					X					X					0				
VER SP 2.3	3	3					X				X	X					0				
VER SP 3.1	5	4					X	X			X	X					1	X			
VER SP 3.2	10	7					X		X			X	X	X	X	X	3	X	X	X	

Table 6.5 Mapping VAL to Both ISO 9001 and ISO 20000

CMMI	Total 9001/ 20000	*ISO 9001* Total	6.3	6.4	7.3.1	7.3.3	7.3.6	7.5.2	7.5.3	7.6	8.2.1	8.2.4	8.4	*ISO 20000* Total	5	10.1
			1	1	1	1	5	5	1	1	1	5	1		3	1
VAL SP 1.1	4	4			X		X	X				X		0		
VAL SP 1.2	6	5	X	X			X	X				X		1		X
VAL SP 1.3	6	5				X	X	X		X		X		1	X	
VAL SP 2.1	6	5					X	X			X	X	X	1	X	
VAL SP 2.2	5	4					X	X	X			X		1	X	

Table 6.6 Mapping MA to Both ISO 9001 and ISO 20000

CMMI	Total 9001/ 20000	*ISO 9001* Total	5.4.1	7.2.2	7.5.3	7.6	8.1	8.2.1	8.2.3	8.2.4	8.4	8.5.1	*ISO 20000* Total	4.3	4.4.3	6.1	6.2	6.3	6.5	6.6	8.3
			1	1	1	4	3	3	4	2	8	2		8	2	2	3	4	5	3	1
MA SP 1.1	9	6	X	X		X		X			X	X	3	X					X	X	
MA SP 1.2	9	5				X	X	X	X		X		4	X				X	X	X	
MA SP 1.3	6	4				X	X		X		X		2	X					X		
MA SP 1.4	8	4				X	X		X		X		4	X			X		X	X	
MA SP 2.1	6	2								X	X		4	X	X			X			X
MA SP 2.2	11	5						X	X	X	X	X	6	X	X	X	X	X	X		
MA SP 2.3	2	1									X		1	X							
MA SP 2.4	6	2			X						X		4	X		X	X	X			

This PA shows great synergy among all frameworks, which will benefit organizations considering their use.

The situation is markedly different for quality assurance, as we can see in Table 6.7.

What is this map telling us? Here we see that ISO 9001 section 8.2.2 "Internal audit" satisfies CMMI and provides a solid infrastructure for ISO 20000-based organizations, since ISO 20000 has no requirement for process evaluation.

The mapping of CM to ISO 9001 and ISO 20000 is shown in Table 6.8.

What is this map telling us? In this mapping we see that CMMI is mapped well to both frameworks. One of the practices, SP 1.2 Establish a Configuration Management System that did not map to ISO 9001 but did map strongly to ISO 20000

Table 6.7 Mapping PPQA to ISO 9001 and ISO 20000

	Total 9001/	*ISO 9001*	*4.2.1*	*8.1*	*8.2.2*	*8.3*	*8.5.2*	*8.5.3*	*ISO 20000*	*3.2*	*4.3*	*4.4.1*
CMMI	20000	Total	1	1	4	1	2	1	Total	1	2	1
PPQA SP 1.1	1	1			X				0			
PPQA SP 1.2	2	2		X	X				0			
PPQA SP 2.1	6	4			X	X	X	X	2		X	X
PPQA SP 2.2	5	3	X		X		X		2	X	X	

Table 6.8 Mapping CM to ISO 9001 and ISO 20000

	Total 9001/	*ISO 9001*	*4.2.3*	*7.3.7*	*7.5.1*	*7.5.3*	*8.2.4*	*8.3*	*ISO 20000*	*5*	*6.3*	*6.4*	*6.6*	*7.2*	*7.3*	*8.3*	*9.1*	*9.2*	*10.1*
CMMI	20000	Total	2	3	1	4	1	2	Total	2	2	1	1	2	2	2	6	3	2
CM SP 1.1	4	2	X			X			2								X	X	
CM SP 1.2	1	0							1								X		
CM SP 1.3	2	1			X				1								X		
CM SP 2.1	11	3		X		X		X	8	X	X	X		X	X	X		X	X
CM SP 2.2	13	4	X	X		X		X	9	X	X		X	X	X	X	X	X	X
CM SP 3.1	3	2		X		X			1								X		
CM SP 3.2	2	1					X		1								X		

provides the process for establishing a system for storing and retrieving configuration items and their archived version, updating CM records, creating reports, and conducting audits. The CM process is necessary for effective and efficient handling of process and product items. The synergy among the frameworks makes their use beneficial to organizations that may consider their joint implementation.

Now, let's review organizational processes. The OPF and OPD maps are shown in Table 6.9 and Table 6.10, respectively.

What are these maps telling us? CMMI and ISO 9001 provide a strong process infrastructure for monitoring implementation and including process improvement experiences in new organizational process assets, which can be used in ISO 20000-based process improvement. However, OPD indicates several major mapping deficiencies for both ISO 9001 and ISO 20000. Organizations using both ISO frameworks would benefit from establishing a measurement repository and process

Table 6.9 Mapping OPF to ISO 9001 and ISO 20000

CMMI	Total 9001/ 20000	*ISO 9001* Total	*4.1* 2	*5.3* 1	*5.4.1* 1	*5.6.1* 2	*8.1* 4	*8.2.2* 6	*8.4* 2	*8.5.1* 9	*8.5.2* 3	*8.5.3* 1	*ISO 20000* Total	*4.3* 1	*4.4.2* 6	*4.4.3* 7	*9.2* 2
OPF SP 1.1	5	4		X	X			X		X			1			X	
OPF SP 1.2	8	5				X	X	X	X	X			3	X	X	X	
OPF SP 1.3	8	5				X	X	X	X	X			3		X	X	X
OPF SP 2.1	7	4					X	X		X	X		3		X	X	X
OPF SP 2.2	6	4	X					X		X	X		2		X	X	
OPF SP 3.1	5	3					X			X	X		2		X	X	
OPF SP 3.2	3	1								X			2		X	X	
OPF SP 3.3	3	3	X					X		X			0				
OPF SP 3.4	2	2								X		X	0				

Table 6.10 Mapping OPD to ISO 9001 and 20000

CMMI	Total 9001/ 20000	*ISO 9001* Total	*4.1* 1	*4.2.1* 3	*4.2.2* 1	*6.1* 1	*6.3* 1	*6.4* 1	*7.1* 3	*ISO 20000* Total	*3.2* 1	*4.4.2* 1	*4.4.3* 1	*7.3* 1	*10.1* 1
OPD SP 1.1	9	4	X	X	X				X	5	X	X	X	X	X
OPD SP 1.2	2	2		X					X	0					
OPD SP 1.3	2	2		X					X	0					
OPD SP 1.4	0	0								0					
OPD SP 1.5	0	0								0					
OPD SP 1.6	3	3				X	X	X		0					
OPD SP 2.1	0	0								0					
OPD SP 2.2	0	0								0					
OPD SP 2.3	0	0								0					

asset library. Neither ISO framework addresses the IPPD addition (OPD SG 2). Therefore, CMMI may be used when establishing Integrated Product Teams (IPTs) as an aid to collaboration and communication among project staff.

In the training area, ISO 20000 would benefit from the infrastructure provided by both CMMI and ISO 9001 shown in Table 6.11.

Table 6.11 Mapping OT to ISO 9001 and 20000

CMMI		Total 9001/ 20000	*ISO 9001* Total	*6.2.1* 7	*6.2.2* 7	*ISO 20000* Total	*3.3* 2
OT	SP 1.1	3	2	X	X	1	X
OT	SP 1.2	2	2	X	X	0	
OT	SP 1.3	2	2	X	X	0	
OT	SP 1.4	2	2	X	X	0	
OT	SP 2.1	3	2	X	X	1	X
OT	SP 2.2	2	2	X	X	0	
OT	SP 2.3	2	2	X	X	0	

CMMI Mapping to ISO 12207 and ISO 15288

We will now consider the CMMI mapping to the ISO 15288 and ISO 12207 frameworks. Because of the size of the joint maps, we will limit our discussion to a few examples. The reader may wish to refer to chapter 5 when comparing the individual maps.

The individual maps of PP to ISO 12207 and ISO 15288 are shown in chapter 5. Comparing those two mapping, we see that PP was mapped very sparsely over a large number of ISO requirements.

What is this map telling us? Project planning is an area that is very well represented in both frameworks. We can see from Tables 5.47, 5.48, and 5.60 in chapter 5 that the planning processes in these ISO standards are spread over a large number of sections. The reason for the sparseness of mapping is due to the much larger granularity of ISO requirements compared to the wording of CMMI specific and generic practices. Take, for example, PP SP 2.1 Establish Budget and Schedule. This SP maps to ISO 12207 requirements 6.3.1.3.2.1 "Prepare plans" (which includes ten tasks) and 6.3.4.3.5.3 "Monitor new risks," and to ISO 15288 requirements 6.3.1.3 a1, b1, b2, b3, and 6.3.4.3 c3. As another example, consider PP SP 2.5, which maps to a single activity with five tasks in ISO 15288 but to two sets of requirements addressing skill identification and skill development in ISO 12207. Going in the other direction we see that ISO 12207 requirement 6.3.1.3.2.1 "Prepare plans" maps to all PP SPs, whereas the ISO 15288 requirements are documented separately. So, be careful when interpreting requirements in these frameworks and be aware that they may actually address the same topics.

Let's consider a few examples of mapping CMMI Engineering PAs to both ISO standards. The RD and REQM PAs are commonly considered together as shown in Table 6.12.

Table 6.12 Mapping RD and REQM to ISO 12207 and ISO 15288

CMMI	Total 12207/15288	Total 12207	Total 15288	ISO 15288: 6.4.1.3 a1 6.4.1.3.1.1	a2 6.4.1.3.2.1	b1 6.4.1.3.2.2	b2 6.4.1.3.2.3	b3 6.4.1.3.2.4	b4 6.4.1.3.2.5	c1 6.4.1.3.3.1	c2 6.4.1.3.4.1	c3 6.4.1.3.4.2	c4 6.4.1.3.4.3	c5 6.4.1.3.5.1	c6 6.4.1.3.5.2	ISO 12207 7.1.2.3.1.1	7.1.2.3.1.2	7.1.3.3.1.1	7.1.6.3.1.5	7.1.7.3.1.1	7.2.4.3.2.1	7.2.4.3.2.2	7.2.4.3.2.3
				1	1	2	1	1	5	1	2	1	1	3	1	5	2	1	2	1	1	1	1
PP SP 2.6	2	1	1	X																			
RD SP 1.1	6	3	3		X	X			X														
RD SP 1.2	6	3	3			X			X					X									
RD SP 2.1	3	2	1						X							X							
RD SP 2.2	3	2	1											X		X							
RD SP 2.3	3	2	1						X							X							
RD SP 3.1	5	3	2				X	X								X							
RD SP 3.2	3	2	1						X							X							
RD SP 3.3	6	4	2							X	X						X				X		
RD SP 3.4	2	1	1								X												
RD SP 3.5	4	2	2									X	X										
REQM SP 1.1	0	0	0																				
REQM SP 1.2	0	0	0																				
REQM SP 1.3	2	1	1											X									
REQM SP 1.4	8	7	1												X		X	X	X	X		X	X
REQM SP 1.5	1	1	0																X				

What is this map telling us? ISO 12207 and ISO 15288 both include section 6.4.1 "Stakeholder Requirements Definition Process" and map identically to CMMI, as shown in the table. However, the additional software-specific section 7.1.2 "Software Requirement Analysis Process" in ISO 12207 provides mappings to many RD and REQM practices. Some additional mapping of ISO 12207 to CMMI, and especially to the bidirectional traceability requirement (REQM SP 1.4) comes from sections 7.1.3 "Software Architectural Design Process," 7.1.6 "Software Integration Process," 7.1.7 "Software Qualification Testing Process," and 7.2.4 "Software Verification Process." Here we see that for software, ISO 12207 may contribute some detailed explanations missing from CMMI and ISO 15288.

If we compare maps of the TS PA to ISO 12207 and ISO 15288, we see that all practices are mapped quite well. The mapping to the PI PA is somewhat different, as shown in Table 6.13.

What is this map telling us? Here we see that ISO 15288 does not map to PI SP 1.3 Establish Product Integration Procedures and Criteria, whereas ISO 12207 provides several mappings that nicely trace integration over the life cycle, namely, specifying test requirements during architectural design (7.1.3.3.1.5), updating those requirements in detailed design (7.1.4.3.1.6) and software construction (7.1.5.3.1.4), and developing qualification tests (7.1.6.3.1.4). On the other hand, ISO 12207 has no mapping to PI SP 1.2 Establish the Product Integration Environment, SG 2 Ensure Interface Compatibility, and SP 3.1 Confirm Readiness of Product Components for Integration. Here we can see how the synergy between CMMI and ISO 15288 helps ISO 12207 address the project integration environment and the interface management needed for successful integration, and ISO 12207 helps ISO 15288 with establishing process improvement procedures and criteria for conducting integration.

Another interesting example is Verification. ISO 12207 has requirements for several verification methods that can be used either separately or in conjunction with one another in sections 7.1.7 "Software Qualification Testing Process," 7.2.3 "Software Quality Assurance Process," 7.2.4 "Software Verification Process," and 7.2.6 "Software Review Process." Its interaction with ISO 15288 is shown in Table 6.14A and Table 6.14B.

What is this map telling us? Here we see that ISO 15288 did not map to VER SP 2.1 and SP 2.3, both of which deal with peer reviews. The synergy between CMMI and ISO 12207 provides those additional practices. Whereas CMMI leaves the timing of verification activities to an organization's process architecture design and project tailoring, ISO 12207 specifies points in the life cycle where the verification process should be invoked, for example, in software requirement analysis (7.1.2), software architectural design (7.1.3), software detailed design (7.1.4), software implementation (7.1.5), and software integration (7.1.6). In this area, ISO 12207 provides better guidance than CMMI. Use of all frameworks will certainly help organizations establish verification processes across the life cycle.

Table 6.13 Mapping PI to ISO 12207 and ISO 15288

																	6.4.5.3							6.4.7.3			
	Total 12207/	ISO 12207	6.1.2.3.5.1	6.1.2.3.6.2	6.4.5.3.1.1	6.4.6.3.1.4	7.1.3.3.1.5	7.1.4.3.1.6	7.1.5.3.1.4	7.1.6.3.1.1	7.1.6.3.1.2	7.1.6.3.1.4	7.1.6.3.1.5	7.2.4.3.2.4	ISO 15288	6.4.4.3 b3	a1	a2	b1	b2	b3	b4	b5	a1	a2	b1	b2
CMMI	15288	Total	1	1	2	1	1	1	2	2	2	1	1	1	Total	1	1	1	1	1	1	2	1	1	1	1	1
PI SP 1.1	4	2							X	X					2		X	X									
PI SP 1.2	1	0													1				X								
PI SP 1.3	5	5					X	X	X	X		X			0												
PI SP 2.1	0	0													0												
PI SP 2.2	1	0													1							X					
PI SP 3.1	2	0													2					X	X						
PI SP 3.2	3	2			X						X				1							X					
PI SP 3.3	5	4			X						X		X	X	1								X				
PI SP 3.4	11	3	X	X		X									8	X								X	X	X	X

Table 6.14A Mapping VER to ISO 12207

CMMI	Total 12207/ 15288	*ISO 12207* Total	*6.4.2.3.2.1*	*6.4.3.3.2.1*	*6.4.5.3.1.1*	*6.4.5.3.2.1*	*6.4.6.3.1.1*	*6.4.6.3.1.2*	*7.1.2.3.1.2*	*7.1.3.3.1.5*	*7.1.3.3.1.6*	*7.1.4.3.1.5*	*7.1.4.3.1.6*	*7.1.4.3.1.7*	*7.1.5.3.1.4*	*7.1.5.3.1.5*	*7.1.6.3.1.5*	*7.1.7.3.1.1*	*7.1.7.3.1.3*	*7.2.1.3.2.3*	*7.2.3.3.2.3*	*7.2.4.3.1.1*	*7.2.4.3.1.4*	*7.2.4.3.1.6*	*7.2.6.3.3.1*
			3	3	1	3	1	1	3	1	3	2	1	2	1	3	8	1	1	3	2	1	1	5	5
VER SP 1.1	7	6										X					X					X	X		
VER SP 1.2	6	4															X								
VER SP 1.3	11	9				X				X		X	X		X		X								
VER SP 2.1	10	10	X	X					X		X					X	X			X				X	X
VER SP 2.2	12	11	X	X					X		X					X	X			X				X	X
VER SP 2.3	10	10	X	X					X		X					X	X			X				X	X
VER SP 3.1	12	11			X	X	X							X			X	X			X			X	X
VER SP 3.2	15	12				X		X						X			X		X		X			X	X

Table 6.14B Mapping VER to ISO 15288

CMMI	Total 12207/ 15288	*ISO 15288* Total	*6.4.4.3 b2* 1	*6.4.6.3 a1* 3	*a2* 1	*a3* 0	*b1* 1	*b2* 2	*b3* 1	*b4* 1
VER SP 1.1	7	1		X						
VER SP 1.2	6	2		X			X			
VER SP 1.3	11	2		X	X					
VER SP 2.1	10	0								
VER SP 2.2	12	1						X		
VER SP 2.3	10	0								
VER SP 3.1	12	1						X		
VER SP 3.2	15	3	X						X	X

CMMI Maps to All Frameworks

When dealing with CMMI maps to all frameworks, it is interesting to explore individual subprocesses to see how multiple frameworks contribute to developing a deeper understanding of that subprocess. Based on the process architecture chosen, a subprocess may address an entire CMMI PA, part of the PA, or a combination of selected goals and practices from several PAs.

The project management process may be decomposed into several subprocesses, one of which is project planning. Project planning can be further decomposed into subprocesses or process elements such as estimation and project plan development. If we decompose the project planning process using, for example, CMMI PP SGs as subprocesses, we end up with process elements for Estimation, Plan Development, and Plan Commitment. We can now examine how the estimation process will be helped by using several ISO standards. The estimation process is defined in CMMI PP SG 1 Establish Estimates, which has four associated SPs, and in IPM SP 1.2 Use Organizational Process Assets for Planning Project Activities. The mapping of those SPs to all four ISO frameworks is shown in Table 6.15, which is a subset of the maps shown in chapter 5. The first column labeled "Total" shows the sum of mappings from each ISO standard and indicates the synergy of each ISO standard paired with CMMI and, in turn, of all five standards. We see that IPM SP 1.2 has no mapping to ISO 20000 or ISO 12207 but is mapped to ISO 9001 and ISO 15288.

What is this map telling us? In CMMI, the PP and IPM PAs deal with estimation. We noted in chapter 3 that IPM is an advanced PA that builds on PP, which is a basic PA. An organization that wants to establish an estimation process may decide to consider not only the PP PA but also the IPM PA to ensure that the

Table 6.15 Estimation Subprocess Mapping

CMMI	Total All Frameworks	ISO 9001	7.1	7.2.1	7.2.2	7.2.3	7.3.1	7.3.2	7.3.3	ISO 20000	4.1	5	6.4	ISO 12207	6.1.2.3.4.2	6.2.3.3.1.3	6.3.1.3.1.1	6.3.1.3.2.1	7.1.1.3.1.1	7.1.1.3.1.3	ISO 15288	6.3.1.3 a1	6.3.1.3 a2	6.3.1.3 a3	6.3.1.3 a4	6.3.1.3 b1	6.3.1.3 b2	6.3.1.3 b3	6.3.2.3 b5
		Total	2	0	0	0	5	1	0	Total	2	3	3	Total	2	1	1	4	2	1	Total	1	1	2	1	2	2	2	1
IPM SP 1.1	11	2	X				X			3	X	X	X	3	X				X	X	3	X		X			X		
IPM SP 1.2	2	1						X		0				0							1							X	
PP SP 1.1	10	1					X			3	X	X	X	3		X	X	X			3		X		X				X
PP SP 1.2	3	1					X			0				1				X			1					X			
PP SP 1.3	7	2	X				X			0				3	X			X	X		2			X			X		
PP SP 1.4	6	1					X			2		X	X	1				X			2					X		X	

organization's measurement repository, containing historical data and associated assumptions and rationale, is used when performing those estimates. The requirements in ISO 15288 section 6.3.1.3 b3 "Define budget" and ISO 9001 section 7.3.2 "Similar designs" contribute to ensuring completeness of the overall estimation process. Although ISO 20000 does not map to PP SP 1.2 Establish Estimates of Work Products and Task Activities or SP 1.3 Define Project Lifecycle, we see that ISO 9001 section 7.3.1, "Design and development," ISO 12207 section 6.3.1.3.2.1 "Prepare plans," and ISO 15288 sections 6.3.1.3 b1 "Define schedule" and b2 "Define criteria" supplement those CMMI practices. There are several important notes in both ISO 12207 and ISO 15288 associated with those requirements that provide additional hints when performing estimation. If the project's domain is information technology, ISO 20000 provides additional requirements related to:

- Service management planning and the scope of service management (section 4.1)
- Ensuring that new or changed services are deliverable and manageable at the agreed cost and schedule (section 5)
- Ensuring that budget and accounting for the cost of service provision is specified (section 6.4)

It also means that if a service management project needs an estimation process, it will have to rely on the support from CMMI and ISO 15288. When comparing ISO frameworks to CMMI, we see that they are more compliance related, whereas CMMI provides more process guidance. Together they help to establish a very strong estimation process that is both complete and answers compliance requirements.

Similarly, the maps in chapter 5 show that there are no PMC SPs that are not mapped to the ISO frameworks and all PMC practices are well represented in all frameworks.

In ISO 12207 we have a case where several requirements reference other sections of the standard, including 7.2.6 "Software Review Process," 7.2.7 "Software Audit Process," and 7.2.8 "Software Problem Resolution Process." Those referenced sections are separately mapped to CMMI and can be efficiently used to create their own process elements, such as a review process or an audit process, with minimal coupling and maximal cohesion, which can then be invoked anywhere in the life cycle.

Now, let's suppose that a process designer would like to develop a QA process satisfying all five frameworks by considering the mapping shown in Table 6.16. Here again the column "Total All Frameworks" contains the sum of all individual framework maps.

What is this map telling us? ISO 20000 has no map to the two PPQA practices dealing with objective evaluation. The ISO 20000 standard requires continual improvement (section 4.4) but, although compliance with the service management plans or the standard itself is required, it never explicitly specifies review of service

Table 6.16 Mapping PPQA to All Frameworks

CMMI	Total All Frameworks	*ISO 9001*	*4.2.1*	*8.1*	*8.2.2*	*8.3*	*8.5.2*	*8.5.3*	*ISO 20000*	*3.2*	*4.3*	*4.4.1*	*ISO 12207*	*6.2.5.3.2.1*	*6.2.5.3.2.2*	*6.3.2.3.3.1*	*7.2.1.3.2.3*	*7.2.3.3.1.4*	*7.2.3.3.1.5*	*7.2.3.3.1.6*	*7.2.3.3.2.1*	*7.2.7.3.1.6*	*7.2.7.3.1.7*	*7.2.7.3.2.1*	*ISO 15288*	*6.2.5.3 c1*	*6.2.5.3 c2*	*6.3.2.3 a2*	*6.3.2.3 b4*	*6.4.4.3 b2*
		Total	1	1	4	1	2	1	Total	1	2	1	Total	1	1	1	1	4	1	2	1	1	1	2	Total	1	1	4	1	1
PPQA SP 1.1	10	1			X				0				8					X		X				X	1			X		
PPQA SP 1.2	15	2		X	X				0				11			X	X	X		X	X			X	2			X		X
PPQA SP 2.1	14	4			X	X	X	X	2		X	X	5	X				X				X	X		3	X		X	X	
PPQA SP 2.2	10	3	X		X		X		2	X	X		3		X			X	X						2		X	X		

management processes. This gap can be addressed by reviewing documents as opposed to reviewing processes. Thus, service management processes would benefit from systematic and periodic reviews by objective reviewers who would evaluate service management processes against criteria to minimize subjectivity and bias.

We could list more individual process examples, but the interested reader can use the additional mapping material available on the publisher's Web site to construct similar process examples for creating process elements and for process improvement.

Now we can address another important issue, namely, the relationship among PAs, their SPs, and GPs. In chapter 5, Figure 5.4, we described the interaction between PAs, GPs, and ISO requirements. Now let's take a look at some examples.

Our first example deals with interaction between PP SP 2.4 Plan for Project Resources and GP 2.3 Provide Resources, in all PAs. Table 6.17 uses framework maps to illustrate how those requirements interact. (Organizational PAs are omitted since their resource requirements differ from the project-related PAs.)

What is this map telling us? PP SP 2.4 is highlighted in Table 6.17. We notice that ISO 9001 does not map to GP 2.3 in the DAR and RSKM PAs. (As we recall from chapter 5, the DAR and RSKM PAs did not map to ISO 9001 at all.) Similarly, ISO 20000 does not map to GP 2.3 in the CAR PA nor in several other weakly mapped PAs. Another thing we can see in that table is that GP 2.3 in all other PAs is supported by ISO 9001 requirements in several sections. Similarly, ISO 12207 and ISO 15288 resource requirements for portfolio management (section 6.2.3) and the ISO 15288 requirement for defining requirements and identifying and providing resources for the infrastructure (section 6.2.2) map to GP 2.3 in all PAs. This last requirement is interesting from the wording point of view since it deals with the project infrastructure even though it is associated with the organizational infrastructure process. PP SP 2.4, which drives all instances of GP 2.3 in project-related PAs, is mapped to a variety of requirements in the ISO standards. These requirements address topics including:

- ISO 9001 sections 6.3 "Resources for providing services and equipment;" 6.4 "Resources for maintaining environment;" 7.3.1 "Update plans"
- ISO 20000 sections 4.1 "Resources, facilities, and budget;" 6.5 "Manage capacity"
- ISO 12207 sections 6.3.1.3.2.1 "Prepare plans;" 7.2.6.3.1.2 "Agree on resources"
- ISO 15288 sections 6.3.1.3 b4, b5, and b6 related to planning project resources

Here we see that the ISO 20000, ISO 12207, and ISO 15288 requirements for resource planning are well aligned with CMMI, whereas ISO 9001 only addresses provision of resources. From those maps we can see that each framework has mappings to GP 2.3 in all pertinent PAs, but the planning component is missing in ISO 9001. Therefore, ISO 9001-registered organizations would benefit from the guidance found in other frameworks when planning and establishing resources.

Table 6.17 Interaction of PP SP 2.4 and GP 2.3 for All Frameworks

CMMI		Total All	*ISO 9001*	*4.1*	*5.1*	*5.6.3*	*6.1*	*6.3*	*6.4*	*7.3.1*	*7.5.1*	*ISO 20000*	*3.1*	*4.1*	*4.2*	*5*	*6.5*	*ISO 12207*	*6.2.3.3.1.4*	*6.3.1.3.2.1*	*6.3.4.3.1.4*	*6.3.7.3.1.6*	*6.3.7.3.1.7*	*7.2.5.3.1.4*	*7.2.6.3.1.2*	*7.2.7.3.1.3*	*ISO 15288*	*6.2.2.3 a1*	*6.2.2.3 a2*	*6.2.3.3 a4*	*6.3.1.3 b4*	*6.3.1.3 b5*	*6.3.1.3 b6*	*6.3.4.3 a4*
		Frameworks	Total	15	15	15	15	16	16	1	2	Total	4	12	12	13	1	Total	17	1	1	1	1	1	2	1	Total	17	18	17	1	2	1	1
CAR	GP 2.3	10	6	X	X	X	X	X	X			0						3	X								3	X	X	X				
CM	GP 2.3	13	6	X	X	X	X	X	X			3		X	X	X		3	X								3	X	X	X				
DAR	GP 2.3	4	0									0						3	X								3	X	X	X				
IPM	GP 2.3	14	6	X	X	X	X	X	X			4	X	X	X	X		3	X								3	X	X	X				
MA	GP 2.3	15	6	X	X	X	X	X	X			3		X	X	X		3	X			X	X				3	X	X	X				
PI	GP 2.3	11	7	X	X	X	X	X	X		X	0						3	X								3	X	X	X				
PMC	GP 2.3	15	6	X	X	X	X	X	X			4	X	X	X	X		3	X						X		3	X	X	X				
PP	SP 2.4	12	3					X	X	X		3		X		X	X	4		X					X		4		X		X	X	X	
PP	GP 2.3	15	6	X	X	X	X	X	X			4	X	X	X	X		4	X								4	X	X	X		X		
PPQA	GP 2.3	15	6	X	X	X	X	X	X			4	X	X	X	X		3	X							X	3	X	X	X				
QPM	GP 2.3	10	6	X	X	X	X	X	X			0						3	X								3	X	X	X				
RD	GP 2.3	13	6	X	X	X	X	X	X			3		X	X	X		3	X								3	X	X	X				
REQM	GP 2.3	13	6	X	X	X	X	X	X			3		X	X	X		3	X								3	X	X	X				
RSKM	GP 2.3	8	0									2			X	X		4	X		X						4	X	X	X				X
SAM	GP 2.3	13	6	X	X	X	X	X	X			3		X	X	X		3	X								3	X	X	X				
TS	GP 2.3	11	7	X	X	X	X	X	X		X	0						3	X								3	X	X	X				
VAL	GP 2.3	14	6	X	X	X	X	X	X			3		X	X	X		3	X					X			3	X	X	X				
VER	GP 2.3	13	6	X	X	X	X	X	X			3		X	X	X		3	X								3	X	X	X				

There are some major deficiencies in satisfying certain GPs. One of the least frequently satisfied generic practices is GP 3.2 Collect Improvement Information, which is related to IPM SP 1.6 Contribute to the Organizational Process Library, and is extremely important for institutionalization. By inspection of the maps, we can see that IPM SP 1.6 is only mapped to ISO 9001 section 8.4 "Collect data on QMS effectiveness," where the emphasis is on process effectiveness rather than on collection of all pertinent data and lessons learned. This means that a major process improvement effort will have to be initiated to ensure that this SP and the associated GP 3.2 are implemented.

What remains to be discussed are the organizational level PAs. When we discuss the organizational PAs, it is almost impossible to address only the SPs since the GPs bring institutionalization into the picture and these PAs are all about the process infrastructure and institutionalization. Because Organizational Process Focus (OPF) is a cornerstone of an organization's process infrastructure, let's first consider the OPF maps shown in Table 6.18. Here, for clarity, we have rolled up the matches for each practice to the ISO standard level.

Table 6.18 Mapping OPF to All Frameworks

CMMI	Total All Frameworks	*ISO 9001* 72	*ISO 20000* 28	*ISO 12207* 29	*ISO 15288* 21
OPF SP 1.1	7	4	1	1	1
OPF SP 1.2	11	5	3	2	1
OPF SP 1.3	10	5	3	1	1
OPF SP 2.1	9	4	3	1	1
OPF SP 2.2	9	4	2	2	1
OPF SP 3.1	5	3	2	0	0
OPF SP 3.2	3	1	2	0	0
OPF SP 3.3	5	3	0	1	1
OPF SP 3.4	7	2	0	4	1
OPF GP 2.1	12	7	2	1	2
OPF GP 2.2	11	3	2	4	2
OPF GP 2.3	11	6	1	1	3
OPF GP 2.4	7	3	1	1	2
OPF GP 2.5	3	2	1	0	0
OPF GP 2.6	3	3	0	0	0
OPF GP 2.7	4	1	1	1	1
OPF GP 2.8	10	4	2	2	2
OPF GP 2.9	7	1	1	4	1
OPF GP 2.10	12	9	0	2	1
OPF GP 3.1	2	1	1	0	0
OPF GP 3.2	2	1	0	1	0

What is this map telling us? First, we notice that by taking all the ISO frameworks together, based on the sum of the individual mappings shown in the "All Frameworks" column, all CMMI SPs and GPs are mapped. However, individual maps to ISO frameworks, with exception of ISO 9001, are less complete. ISO 9001 provides a great foundation for continual process improvement. However, the other three ISO standards have several deficiencies in this area. ISO 20000 provides some guidance for determining and planning process improvements, but falls short when those plans have to be monitored and process experiences have to be documented. Here we also see that the GPs that are important for establishing the process infrastructure are well represented in ISO 9001, but are weakly mapped in the other ISO frameworks. With the synergy that can be seen in the map, a process architect can use CMMI and ISO 9001 to provide a solid process foundation, which can be extended by the other three frameworks.

A similar situation is seen for the OPD PA as shown in Table 6.19.

Table 6.19 Mapping OPD to All Frameworks

CMMI		Total All Frameworks	*ISO 9001*	*ISO 20000*	*ISO 12207*	*ISO 15288*
			47	9	35	25
OPD	SP 1.1	16	4	5	4	3
OPD	SP 1.2	5	2	0	1	2
OPD	SP 1.3	3	2	0	0	1
OPD	SP 1.4	3	0	0	3	0
OPD	SP 1.5	4	0	0	4	0
OPD	SP 1.6	10	3	0	4	3
OPD	SP 2.1	1	0	0	1	0
OPD	SP 2.2	1	0	0	1	0
OPD	SP 2.3	0	0	0	0	0
OPD	GP 2.1	8	5	0	1	2
OPD	GP 2.2	11	2	0	7	2
OPD	GP 2.3	10	6	0	1	3
OPD	GP 2.4	7	2	1	1	3
OPD	GP 2.5	3	2	1	0	0
OPD	GP 2.6	4	4	0	0	0
OPD	GP 2.7	3	1	0	1	1
OPD	GP 2.8	10	4	1	2	3
OPD	GP 2.9	4	1	0	2	1
OPD	GP 2.10	10	7	1	1	1
OPD	GP 3.1	1	1	0	0	0
OPD	GP 3.2	2	1	0	1	0

What is this map telling us? Here we see somewhat stronger synergy among all frameworks. No single framework predominates and supplies most of the mapping. This may be expected because each of the ISO standards has requirements for establishing a process to guide all projects. They differ in institutionalization, although ISO 9001 provides the strongest map and ISO 20000 the weakest. That means that a process architect would want to use CMMI and ISO 9001 when creating the organizational set of standard processes to be institutionalized. Under OPD SG 2, which addresses IPPD, only one practice, OPD SP 2.3 Balance Team and Home Organization Responsibilities, is not mapped in any ISO framework. It is interesting to note that to establish an organizational measurement repository and process asset library, a process architect would rely on CMMI since neither ISO 9001 nor ISO 20000 have such requirements, whereas ISO 12207 and ISO 15288 have very high level requirements, such as requiring an infrastructure that supports project resource requirements.

Finally, let's consider OT and its mapping to all frameworks, as shown in Table 6.20.

What is this map telling us? We have included two practices, PP SP 2.5 Plan for Needed Knowledge and Skills and PMC SP 1.1 Monitor Project Planning Parameters, in this map since they provide the project-level aspects of training. In Table 6.20 we again see how synergy works in planning, monitoring, and implementing training across the organization. Here we also see that ISO 9001 provides a very strong foundation for establishing organizational training. Both ISO 12207 and ISO 15288 follow with the Human Resource Management Process (section 6.2.3), whereas ISO 20000 lags in requiring establishment of training records and measuring training effectiveness. It is notable that evaluation of training effectiveness is only required by ISO 9001.

Table 6.20 Mapping to All Frameworks

CMMI		Total All Frameworks	*ISO 9001*	*ISO 20000*	*ISO 12207*	*ISO 15288*
			16	11	16	20
OT	SP 1.1	10	2	2	3	3
OT	SP 1.2	3	2	1	0	0
OT	SP 1.3	8	2	1	1	4
OT	SP 1.4	5	2	0	1	2
OT	SP 2.1	6	2	1	1	2
OT	SP 2.2	4	2	0	1	1
OT	SP 2.3	2	2	0	0	0
PMC	SP 1.1	15	0	4	4	7
PP	SP 2.5	10	2	2	5	1

Relationship between ISO 9001 and ISO 20000

Using the mapping data in our database, we created a map between ISO 9001 and ISO 20000 using CMMI as a bridge. This process is shown in Figure 6.2. We reviewed and adjusted the maps to take advantage of our understanding of both ISO 9001 and ISO 20000 and constructed a matrix showing the ISO 9001-to-ISO 20000 relationships similar to the matrices used for other framework relationships. It is important to remember that the maps between CMMI and each of the ISO standards are subjective. The mechanical construction of a second generation map is therefore even more dependent on an understanding of the relevant frameworks.

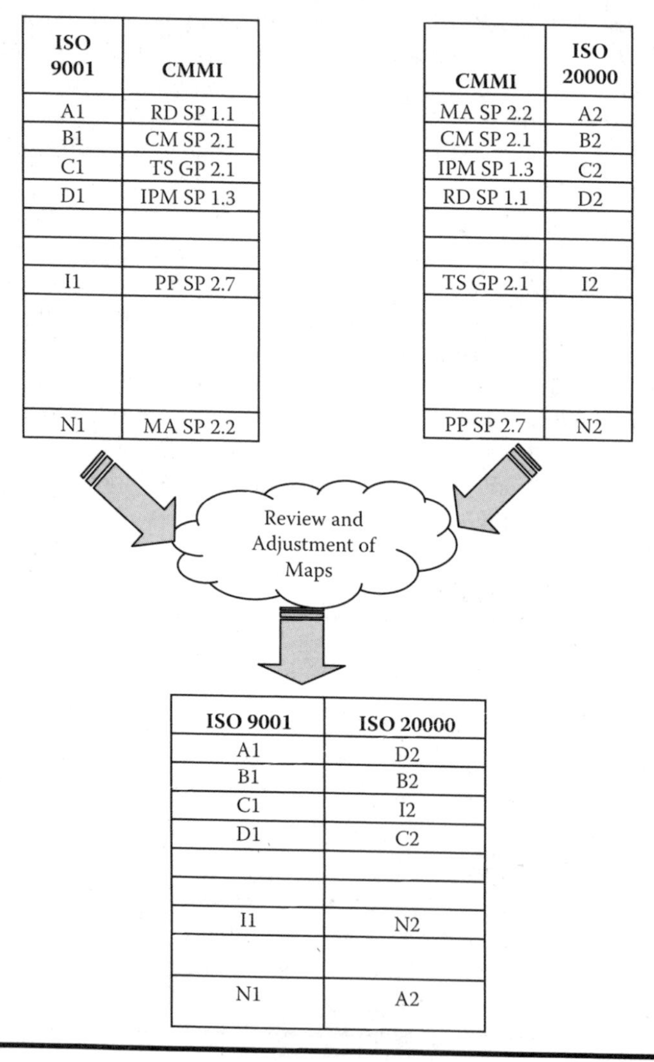

Figure 6.2 Mapping ISO 9001 and ISO 20000 via CMMI.

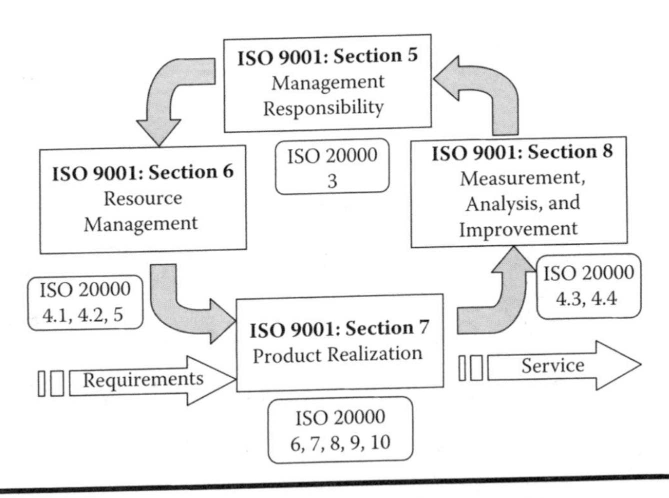

Figure 6.3 Relationship between ISO 9001 and ISO 20000.

Analysis of the resultant matrix revealed several important points. Although both frameworks follow the Shewhart PDCA cycle, the first observation one can make is that those two ISO standards are not well aligned. That is, the requirements of one framework map to several requirements in the other framework. The good news is that the frameworks are compatible, enabling organizations that use both frameworks to capitalize on their synergy. The two frameworks differ in the domains addressed (generic requirements of ISO 9001 vs. service management requirements of ISO 20000) and in the level of requirements detail. In general, ISO 9001 is more detailed than ISO 20000, although the ISO 20000 standard provides more details in specific service management requirements.

If we can equate the quality management processes set forth in ISO 9001 with the service management processes in ISO 20000, we can show a much stronger relationship between those two frameworks. The quality plan can be equated to the service management plan, quality objectives to the service objectives and requirements, and so forth. Figure 6.3 shows how those two frameworks compare.

This relationship is confirmed by the mapping. Because the mapping shows confidence levels and is more detailed and precise than the relationship matrix, several differences are revealed. For example, ISO 9001 section 4.1 specifies major requirements for the quality management system, documentation, the quality manual, and control of documents. Those requirements are spread over several sections in ISO 20000 but there is nothing in that standard about identifying processes since those processes are specifically defined in sections 6 through 10. There are only two ISO 9001 requirements that are not mapped to any ISO 20000 statement:

- Quality management system planning (5.4.2)
- Preservation of product (7.5.5)

Table 6.21 Summary View of the ISO 9001-to-ISO 20000 Mapping

ISO 9001 Section	ISO 20000 Section 3	4	5	6	7	8	9	10
4	Full match	Full match	Strong	Strong	Strong	Very weak	Full match	Full match
5	Medium	Strong	Medium	Weak	Very weak	No match	Very weak	No match
6	Full match	Weak	Full match	Very weak	No match	No match	No match	Weak
7	Very weak	Weak	Medium	Medium	Medium	Very weak	Very weak	Very weak
8	Very weak	Full match	Strong	Strong	Weak	Weak	Very weak	Very weak

It is somewhat surprising that ISO 20000 does not require a QMS. On the other hand, it is no surprise that the preservation of product requirement is not mapped since the product of ISO 20000 *is* service. Specifically, ISO 9001 section 7 "Product realization" should be enhanced by ISO 20000 sections 6 through 10. A high-level view of the relationship between the two standards may be seen in Table 6.21. This table shows that there are several sections that map rather well, although there are also several sections that did not map at all. Remember that rolling up the mapping results masks the details that should be considered when developing a process improvement or appraisal strategy.

Since the Quality Manual is a central document in the ISO documentation hierarchy, we will show how it can be written to address ISO 20000 requirements. The Quality Manual, however, is unique to each organization and the following section is just an example. Typically, there are two ways to approach the design of the Quality Manual: (1) as a stand-alone document that deals only with policy, scope, justified exclusions, the interaction between processes, and references to procedures; and (2) as an integrated document that contains policy, justified exclusions, the interaction between processes, and the procedures (Shlickman 2003).

The first option is more flexible for most organizations, enabling documentation design to target specific audiences, such as senior management, program managers, project managers, engineering managers, process engineers, designers, developers, and members of quality assurance and configuration management functions, who may require different documentation based on their assignment and skill level. Table 6.22 shows a high-level Quality Manual table of contents, based on ISO/TR 10013:2001 (ISO 2001) using the first option.

Table 6.22 Typical Quality Manual Table of Contents

Section	*Title*	*Contents*	*Extensions for ISO 20000*
1	Scope	The scope should define the organization to which the manual applies.	A brief paragraph should describe the documentation hierarchy and its relationship to the service management.
2	Review, Approval, and Revision	Evidence of review, approval, revision status, and date of the Quality Manual should be clearly indicated.	
3	Quality Policy and Quality Objectives	Quality policy and quality objectives can be included in the Quality Manual, or it can be a separate document. The quality goals (measurable entities, which will satisfy the objectives) may be included in the Quality Manual or in the other QMS documents.	Indicate documented service management policies and procedures as well as service management objectives.
4	Organization, Responsibility, and Authority	The structure of the organization should be identified. The responsibility, authority, and interactions may be indicated, for example, by means of organization charts or job descriptions.	In addition to the organization charts, responsibilities covered under section 3.1 may be referenced here.
5	References	List of documents referenced in the text.	
6	Quality Management System Description	The description of the QMS and its implementation in the organization is provided. Here the processes and their interactions are described or referenced. This section should also contain descriptions of, or references to, the documented procedures. *Note:* To facilitate reviews and audits, most QMs list the processes, or references to those processes, in the order of the ISO standard sections.	This section should refer to the service management process, procedures, standards, and methods. This section should also contain the ISO-required processes not addressed by ISO 20000. The Service Management Plan can be cross-referenced here if it provides end-to-end service management process description.
7	Appendices	Any supporting material may be included.	Supporting material (e.g., ISO 9001-to-ISO 20000 map).

Summary

In our discussion of the process improvement aspects of mapping, we described a CMMI-centric approach that accommodates all of the selected ISO standards. We pointed out that the ISO standards are more compliance related, whereas CMMI provides more process guidance. Using the CMMI-centric approach, we provided several examples showing process elements that are based on the use of CMMI and all ISO standards, and indicated where the synergy among the frameworks is used to overcome structural differences between the frameworks. Using the maps created and described in chapter 5, we created a process view that is based on CMMI and multiple ISO standards.

When creating processes, the granularity of framework requirements must be considered. ISO 9001 and ISO 20000 are at a high level, listing fewer but more general requirements, whereas ISO 12207 and ISO 15288 provide specific requirements that can be used in a process architecture as best practices.

Through the use of CMMI mappings, we have also created an ISO 9001-to-ISO 20000 map that may be used for those organizations that plan to use those standards without benefit of CMMI-based process improvement.

Chapter 7

Appraisals

Background

In previous chapters we saw that there is considerable synergy among the selected frameworks. The question is: Can this synergy be used not only for process improvement but also in appraisals? The answer is yes, although there is no single formal approach to address that goal. Most organizations strive to capitalize on framework synergy, and when they successfully implement such an approach, are reluctant to share information about the techniques employed.

ISO registration* audits and the SCAMPI method use different approaches for achieving quite similar goals, namely, ensuring that the particular standard is followed and identifying discrepancies. Whereas ISO audits require discrepancies to be tracked, resolved, and closed, CMMI-based appraisals end with the identification of weaknesses leaving it as the organization's prerogative to decide how to address those weaknesses.

An effective approach to addressing different appraisal methods is to develop a body of objective evidence that simultaneously satisfies each of the selected frameworks. For that purpose, one has to analyze requirements of each appraisal method and then develop an approach for collecting evidence while implementing and executing the organization's and project's processes, as shown in Figure 7.1. In this figure we can see that the implemented process is based on the synergy among several frameworks, each of which have unique audit or appraisal procedures. Those

* In this book, when we discuss ISO audits, we use the terms *registration* and *certification* interchangeably.

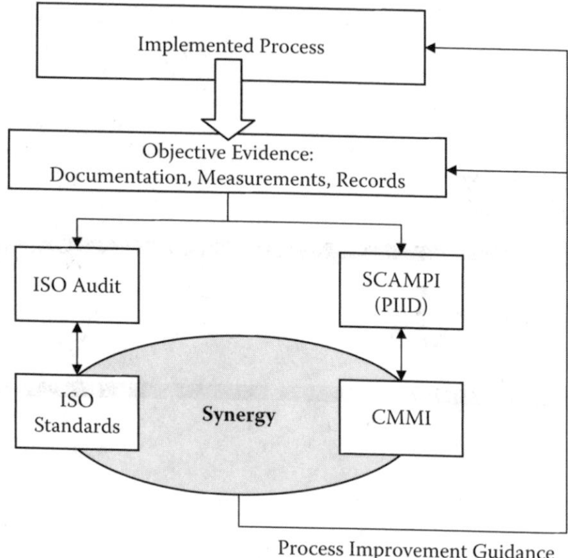

Figure 7.1 Common objective evidence approach.

procedures depend on the collection of objective evidence produced by the implemented process, which, in turn, reflects the framework requirements. As one would expect, the synergy among the frameworks is seen in the architecture of the implemented process. Synergy is also found in the objective evidence since the process elements, common to multiple frameworks, will create evidence that can be used to show successful implementation of those frameworks.

In this chapter we will first discuss the assessment methods associated with each framework, look at their commonalities, and then describe a potential solution that can be implemented to satisfy most methods at the same time.

Characteristics of SCAMPI

SCAMPI stands for Standard CMMI Appraisal Method for Process Improvement. The requirements for the method are captured in the Appraisal Requirements for CMMI (ARC; SCAMPI 2006a) document, whereas the method itself, based on the ARC, is described in the Method Definition Document (MDD; SCAMPI 2006b). The ARC outlines three distinct appraisal classes, which together form a "family" of appraisals. The characteristics of each appraisal class are shown in Table 7.1. An organization will generally find it beneficial to implement these appraisals sequentially, starting with the least costly and lowest impact Class C appraisals and progressing through Class B to Class A.

Table 7.1 Characteristics of CMMI Appraisal Classes

	Class A	*Class B*	*Class C*
Purpose	Evaluate process institutionalization; very rigorous	Evaluate process implementation; assess risk in satisfying goals	Evaluate process development and the process improvement approach
Rating issued	Yes	No	No
Organizational unit coverage	Required	Not required	Not required
Leader requirements	Authorized by the SEI (must use SCAMPI A)	Authorized by the SEI (for SCAMPI B) or person trained and experienced	Authorized by the SEI (for SCAMPI C) or person trained and experienced
Team requirements	4 to 9 members (including lead)	Minimum of 2	Minimum of 1
Evidence	Rigorous; collected at the CMMI practice level; two pieces of evidence required for each practice (documentation and interviews); face-to-face interviews required	Less rigorous than Class A; collected at the CMMI practice level (documentation and interviews)	Documentation or interviews

Any organization may develop its own appraisal method and may elect to show that it is compliant with the ARC, but the SCAMPI A method is the only method that satisfies the requirements for a Class A appraisal. SCAMPI results are reported to the SEI, which is the "steward" of the method, where they are subject to SEI quality assurance review.

We will limit our discussion to SCAMPI A since it is the method that has the most synergy with the ISO-related certification schemes. There are a large number of books that cover CMMI appraisals, such as those by Ahern (2005) and Bush and Dunaway (2005). Therefore, we will not describe the appraisal method at a great length and will limit our SCAMPI discussion to the level needed for the reader to understand the relationship of SCAMPI to other appraisal methods.

The SCAMPI family of appraisals is based on a set of principles that have proven essential over the years, such as management commitment, use of a documented appraisal plan, maintenance of confidentiality and nonattribution, collection of objective evidence, identification of weaknesses, and reporting of the results.

The objectives of the SCAMPI A appraisal are:

- Determine compliance of an organization's processes with CMMI
- Gain insight into the organization's engineering capability by identifying the strengths and weaknesses of its current processes
- Rate the organization's compliance by maturity or capability levels
- Identify development and acquisition risks relative to capability or maturity determinations
- Compare process improvement achievements with the other organizations in the industry

SCAMPI A has three phases: (1) plan and prepare, (2) conduct appraisal, and (3) report results, as shown in Figure 7.2. In the first phase, a plan for the appraisal

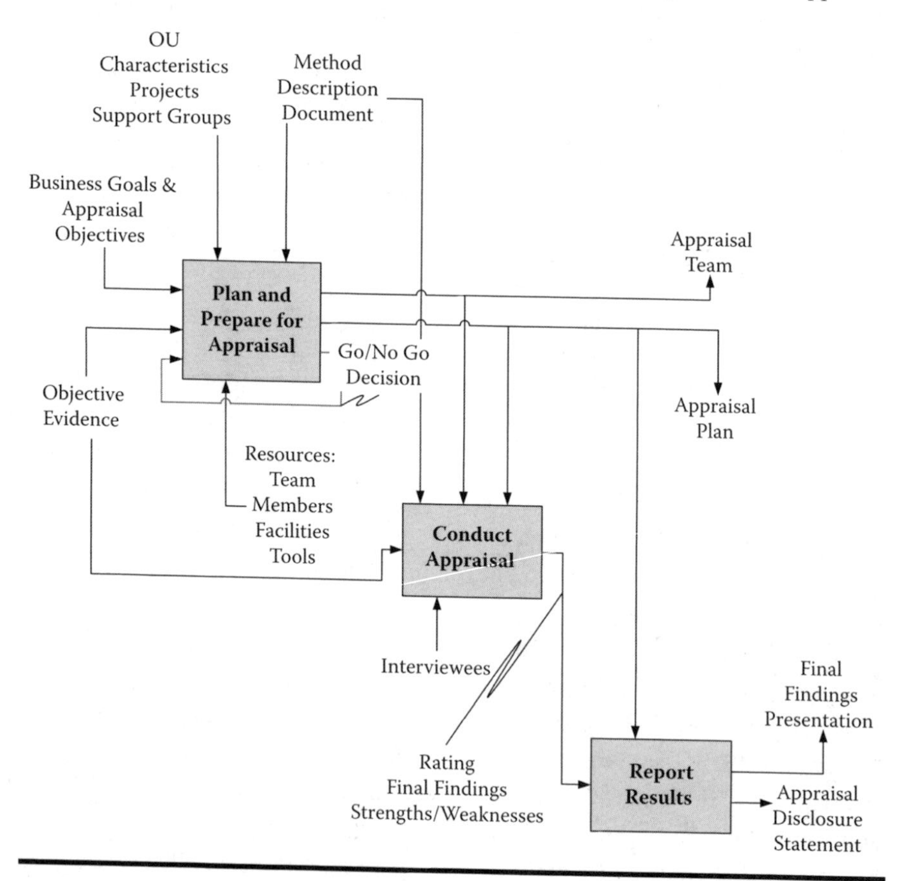

Figure 7.2 SCAMPI A activity flow.

is developed, the appraisal team is selected and trained, evidence is collected and assessed for completeness, and the organization's readiness for the appraisal is determined. The appraisal plan contains several items that require some discussion. The plan is based on appraisal objectives that are derived from the organization's business goals. It may be OK to set achievement of a maturity level as a business objective—and in many cases that is truly a meaningful goal—but there usually are other business and process improvement objectives and constraints, such as reducing the number of delivered defects, reducing delivery time, or increasing productivity.

Appraisal plans may be constrained, for example, by the organization's size (too large to appraise all at once vs. too small to dedicate unlimited resources), expenses, or availability of objective evidence. When the goals and constraints are known, the organization, with help from the lead appraiser (LA), will determine the scope of the appraisal. The appraisal scope has two flavors: the CMMI scope and the organizational scope.

The CMMI scope of the appraisal will be noted in the appraisal plan and indicates the selected CMMI representation, target maturity level or capability profile, PAs to be appraised, and PAs that will be excluded from the appraisal scope.

Determining organizational scope is more complex. First, the organizational unit (OU) that will be subject to the appraisal must be identified. The OU can be the whole organization or, in a large organization, can be defined as a part of the whole organization, or even a single project. Next, this OU will select sample projects and support groups, which will provide evidence of process implementation. Sample projects will provide evidence of process implementation across the project life cycle and support groups will provide evidence of process implementation in the organization's infrastructure and support functions, such as quality assurance, training, and process improvement. Based on the selected project sample and collected evidence, the appraisal team will be able to conclude the extent of process implementation across the organization and its conformance to CMMI. Therefore, it is extremely important that the projects and specific functions selected truly represent the OU. The "critical parameters" that characterize the OU must also be documented in the plan and will typically include:

- The number and size of the projects and support groups across the OU compared to those projects selected for the appraisal
- Number of people in the OU versus those working on the appraised projects
- Representation of different application domains relevant to the organization's work
- Geographical distribution of the projects representing the OU
- Disciplines to be considered, such as systems engineering, software engineering, or manufacturing
- Project types such as legacy versus new development
- Effort type such as maintenance, development, or service
- Customer types, such as commercial versus government
- Life cycle models used in the OU

If an OU contains more than three projects, then, according to the MDD, its scope must "generate at least three instances of each practice in each project-related PA in the model scope of the appraisal." This means that for each specific and generic practice for the PAs in the CMMI scope of the appraisal there must be at least three sets of objective evidence gathered from the projects in the scope of the appraisal. An additional constraint is that the there must be at least one project, called a focus project, that provides evidence for each PA in the selected CMMI scope. Other projects, called nonfocus projects, may provide evidence that complements the evidence of the focus project and satisfies the requirement to have three instances of objective evidence for each practice.

As an example, let's consider an OU with five projects that is targeting CMMI maturity level 3. It selected two focus projects (Projects A and B) and two nonfocus projects (Projects C and D) for the appraisal. Their contribution to the appraisal is shown in Table 7.2.

Note that OPF, OPD, and OT are not project related and therefore do not require the OU to provide three instances of evidence for those PAs.

In the second phase of the appraisal (conduct appraisal), the appraisal team verifies the objective evidence collected by the organization. As explained, each project in the scope of the appraisal collects objective evidence for each SP and GP for those PAs in the scope of the appraisal. It should be noted that the appraisal

Table 7.2 Example of OU Scope

	Project A	*Project B*	*Project C*	*Project D*
REQM	X	X	X	
PP	X	X		X
PMC	X	X		X
SAM	X	X		X
MA	X	X		X
PPQA	X	X		X
CM	X	X		X
OPF				
OPD				
OT				
IPM	X	X		X
RSKM	X	X		X
RD	X	X	X	
TS	X	X	X	
PI	X	X	X	
VER	X	X	X	
VAL	X	X	X	
DAR	X	X	X	

team is looking for the existence of the evidence and its adequacy, not for the subjective "goodness" or quality of provided evidence. Based on this evidence, CMMI practices are characterized at the project level as Fully Implemented, Largely Implemented, Partially Implemented, or Not Implemented. While characterizing CMMI practices, the appraisal team verifies that project processes are implemented and institutionalized. They are obligated to seek both written and oral evidence to determine the extent of the practice conformance with the model. There may be instances where there is insufficient evidence to show that underlying processes are satisfactorily implemented. The appraisal team then documents a "weakness," which is a statement that describes that lack of implementation or institutionalization evidence. Upon completion of the project practice characterization, the results are then rolled up to the organizational level. This is done by combining the characterization of the individual project practices using a fairly simple but repeatable and rigorous rule backed up by appraisal team consensus. Based on the rolled up characterization, satisfaction of the PA goals are rated, leading to determination of PA satisfaction and a maturity-level rating. A goal is rated satisfied if all the associated practices are either Fully Implemented or Largely Implemented and the aggregation of any noted weaknesses does not have a significant negative impact on goal achievement. A PA is satisfied only if all its goals are satisfied. Otherwise, it is declared unsatisfied.

For the staged representation, a maturity level is achieved if all PAs at that level and all the levels below are satisfied (or determined to be not applicable). For ML 3 or higher, there is the additional requirement that GG 3 for each ML 2 PA must also be rated satisfied, meaning that the organization and its projects must revisit ML 2 PAs and specifically ensure that GG 3 is satisfied. For the continuous representation, individual PAs are rated and a PA profile is generated depicting the capability level rating of each PA in the scope of the appraisal.

In the third phase of the appraisal (report results), the LA presents the results to the appraisal sponsor and the appraisal participants. It is important to note that the appraisal team signs the Final Findings Presentation indicating its full agreement with the presented results. The LA completes the Appraisal Disclosure Statement (ADS), which contains the appraisal scope description and indicates the appraisal results (e.g., rating), and submits the appraisal results and the ADS to the SEI for a quality check. When the SEI accepts the appraisal results, the organization may use the rating in its press releases or advertisements. Appraisal results are valid for three years. No surveillance appraisals are required, but most organizations perform internal appraisals to ensure that processes are implemented and sustained.

High maturity appraisals (those that result in CL/ML 4 and 5 ratings) require an SEI-certified high maturity lead appraiser. The ADS for high-maturity appraisals will contain some additional specific information about the scope and conduct of that appraisal.

Characteristics of ISO 9001 Audits

The objectives of ISO audits are to:

- Determine conformance* to ISO 9001:2000, ensuring that specific ISO requirements are satisfied
- Determine Quality Management System (QMS) effectiveness (e.g., for increasing customer satisfaction)
- Evaluate continual improvement of the QMS

Thus, an ISO audit has to answer the following questions:

- Are the management system processes identified and documented?
- Are those processes implemented?
- Are those processes effective in practice and are they continually improved?

Guidelines for QMS auditing are specified in ISO 19011:2002 (ISO 2002), which replaces an earlier ISO 10011 series of standards—ISO 14010, ISO 14011, and ISO 14012—and is applicable to both quality and environmental management systems auditing. In cases where quality and environmental systems are both implemented, they can be audited together. The guidance found in the standard can also be applied to other types of management systems audits. In addition, the ISO 9001 Auditing Practices Group, an informal association of quality management experts, auditors, and practitioners, has developed a set of nonbinding guidelines for auditing QMS that can be found on the International Accreditation Forum (IAF) Web page. The guidelines are very informative, although they have not been endorsed by ISO, its Technical Committee 176, or the IAF.

The ISO audit is based on several important principles, namely, ethical conduct, fair presentation, and due professional care. Those principles ensure that:

- Audits are performed based on trust, integrity, and confidentiality
- Audits are performed according to plan and that results are accurately reported
- Auditors are competent and use proper judgment in reaching their conclusions
- Auditors are independent and free from conflict of interest
- Auditors are objective and reach conclusions based on the presented evidence

The flow of activities for an ISO 9001 audit shown in Figure 7.3 has two stages: preparation and certification. It is based on evaluation of the organization's management processes and conformity of those processes to the standard. Information that is typically collected during audits includes process documentation, review of

---

* *Conformance* or *conformity* is defined by ISO 9000:2000 as "fulfillment of a requirement," whereas *compliance* means conforming to a specific standard or clearly defined law.

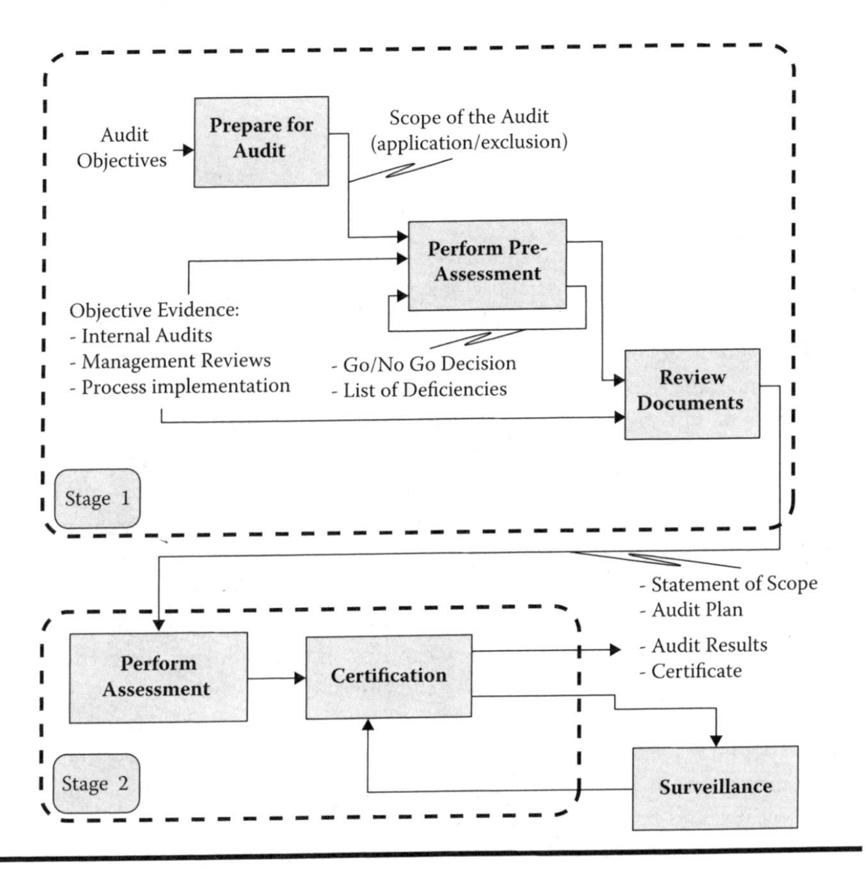

Figure 7.3 ISO 9001 audit activity flow.

records such as results of measurements, records of program monitoring, customer surveys, interviews with employees and management, and observation of activities.

During the preparation stage, the auditor and the audit sponsor must agree on the scope of the audit, which is then documented in the audit plan. Typically, the scope of the audit is the same as the scope of the QMS. The scope should clearly indicate which parts of the organizations and which processes are included in the audit and will receive certification and those that will be excluded. It should list the elements of the organization such as sites, departments, and divisions that are subject to audit and the processes required for product realization or service delivery. The scope must be defined prior to applying for registration or certification and it should be reflected in the QMS and other publicly available documents. In addition to specifying the scope, the audit plan typically addresses:

- Audit objectives
- Audit logistics, such as locations, duration, and travel
- Resources needed

- Identification of the organization's representatives who will be participating in the audit
- Confidentiality
- Potential follow-up actions
- Size and competence of the audit team

The audit evaluates the organizational objectives, which must address the customer needs and market expectations. Those objectives are used to guide the organization to improve process effectiveness, efficiency, and performance. The auditor makes sure that all requirements of ISO 9001 section 4.1—which is used to identify processes, their sequence, the need for process monitoring, resources, and the need for assessing the process effectiveness—are satisfied. This section also applies to outsourced processes, which may affect the product's conformity to requirements. Exclusions to conformity to the standard, which are limited to section 7 "Product realization," must be clearly stated. However, those exclusions should not affect the organization's ability and responsibility to satisfy customer or regulatory requirements for the product. The documentation review is also performed in this phase to enable the auditor to determine if the management system, as described, conforms to the audit criteria. In some cases the documentation review can be conducted during the certification stage of the audit or during a preliminary site visit, if convenient.

During the second stage of the audit—certification—auditors are required to seek and obtain evidence of senior management involvement in the process, including their commitment to continually improving processes, developing policies and process objectives, and reviewing processes to ensure that they are effective and lead to customer satisfaction. The auditors have to investigate implementation of the quality policy and quality objectives, their coherence, and their alignment with the organization's business objectives and customer expectations. In addition, auditors are required to audit the competence of the organization's staff and effectiveness of actions taken.

Auditors are encouraged to use audit checklists. Checklists should be developed for a specific type of audit and used as a planning tool to support consistency of the audits and assist the auditor in collecting notes generated during audits.

ISO does not specify the number of evidence samples required during an audit. It uses the word *adequate* to describe the evidence that gives sufficient confidence that the QMS is implemented as described and satisfies the ISO requirements. Adequacy will, in general, depend on the complexity of the process being audited.

In addition to identifying and documenting nonconformities during the course of the audit, the auditors verify QMS conformity with ISO 9001 requirements. Each nonconformity must be based on evidence that a given requirement was not satisfied and must be clearly described, relating the nonconformity to the ISO requirement. The accuracy and precision of the statement are important because they will be input to the corrective action plans and possibly to the organization's future process improvement plan. The audit conclusions, which are agreed to by the team include:

- The list of nonconformities
- Extent of conformity of the management system with the audit criteria
- Assessment of the management system's capability, effectiveness, and improvement opportunities

Audit records, which include the audit plan, audit report, and nonconformity report, must be maintained and safeguarded, and any other audit documents may be destroyed. The audit team should obey strict confidentiality of the audit results and should not disclose its content without specific permission of the audit sponsor.

Characteristics of ISO 20000 Audits

Because ISO 20000 is a service management system standard, and not a product or a service standard, its audit scope must cover all of the management processes required by the standard. Audit requirements, similar to those of ISO 9001, have been created for ISO 20000 and are governed by the Information Technology Service Management Forum (*it*SMF) Certification Scheme. *it*SMF is a global organization responsible for advancing IT best practices through the use of the IT Infrastructure Library (ITIL). Formal certification against ISO 20000 requires an auditor who is recognized by the Registered Certification Body (RCB), is accredited in accordance with EN 45012:1998 "General requirements for bodies operating assessment and certification/registration of quality systems" and is not at the same time a consultant to the audited organization. (EN 45012:1998 is being replaced by ISO/IEC 17021:2006, *Conformity assessment – Requirements for bodies providing audit and certification/registration of management systems*, effective 15 September 2008.) The organization will, in agreement with the auditor, need to determine and document the scope of the audit in the audit plan. The certification scheme deals with the independence of auditors, conduct of the audit, eligibility, and the scoping requirements for both the standard and organization that will be subject to the audit. Let's take a look at some of those requirements.

The *it*SMF Certification Scheme closely follows the ISO audit flow shown in Figure 7.3. An organization seeking certification under the *it*SMF scheme has to show that it has management control over the processes listed in ISO 20000. "Management control" is defined as:

- Knowledge and control of inputs and outputs
- Definition of measurements
- Objective evidence of process performance and process conformance to ISO 20000 requirements
- Implementation of continual process improvement

The scheme contains specific requirements for clearly, unambiguously, and accurately documenting the audit scope, which is then described in the certificate. The scope of the audit and the certification can be the whole organization or a part of an organization. The scope should also address the geographical distribution of the organization, such as location of the operations that are being audited and the aspects of service involved, such as single services or groups of services and service catalogues. In addition, the ISO 20000 requirements that may not apply have to be identified and documented in the audit plan. The audit scope also typically identifies the organizational or functional boundaries and any outsourced processes.

For audit purposes, evidence of process implementation may come from a single service provider or from the service provider and its suppliers. The scope must include all processes required by the ISO standard. The service provider has to show evidence that it has management control of those processes and that all "shall" statements in the standard are satisfied. Typically, the scope of service management is defined in the service management plan, which then determines the scope of the audit and indicates potential limitation to a single location or a single service.

Although the scope must include the whole management system, it may not include all geographical sites, all data centers, or all customers. Where services are outsourced, the organization will have to show that it controls the interfaces with the outsourced processes or its suppliers. In addition, the organization must provide evidence of controlling improvements of those outsourced service processes. The infrastructure, service desk, and similar functions may be excluded from the audit scope. The evidence provided should make it clear that all service management processes are fully integrated and that the process description covers all service management processes. Process functions may be implemented by different clients.

The organization is required to provide evidence of use of the standard. The audit concentrates on this evidence through document review and interviews of the staff. Typically, three to six months worth of evidence is considered to be adequate proof of implementation of the standard. The certificates are valid for three years after which a re-audit is required. Surveillance audits must be performed at least once per year, whereas internal audits must be performed at regular intervals. During re-audits, the scope of the certification is revisited to ensure that it accurately describes the audit scope.

Similarities and Differences among Audit Approaches

All of the audit approaches discussed here (SCAMPI for CMMI, ISO audits for ISO 9001 and ISO 20000) are based on similar principles:

- Management commitment and definition of audit objectives
- Precise definition of the audit scope
- Development and documentation of an audit plan
- Confidentiality and nonattribution
- Collection of objective evidence based on several sources that describes implementation and improvement of required processes
- Identification of noncompliances
- Reporting of results

ISO 9001 may be applied to organizations in different sectors and industries, whereas ISO 20000 is specific to IT service management, addressing only IT service management processes and their supporting management systems. SCAMPI can only be used in conjunction with CMMI and applies to those organizations that base their process improvement efforts on CMMI. Because ISO 20000 is closely related and complementary to ISO 9001 and some ISO 20000 requirements may be already satisfied by ISO 9001, the scope of an ISO 20000 audit can be defined in such a way to make it coincide with the scope of ISO 9001.

If we look at the objectives of an ISO audit and a SCAMPI appraisal, we see that appraisals will cover most, but not all, aspects of all the frameworks, due to the differences in the structure of the frameworks. For example, SCAMPI will not address topics such as finance or purchasing, and an ISO 9001 audit will not address risk management or the use of a process asset library and measurement repository.

Both assessment methods require an authorized lead who is responsible for planning, conducting, and reporting the results of the assessment. Both methods require that the lead and the sponsor determine the scope of the assessment and that a thorough examination of the objective evidence is performed before a recommendation for certification or a rating can be made.

However, the approaches used to conduct the audit and gather objective evidence for the audits are quite different. SCAMPI has a well-documented method description that lead appraisers and appraisal teams are required to follow, whereas ISO auditors are expected to follow ISO 19011:2002 or the *it*SMF Certification Scheme, both of which are essentially sets of guidelines. Auditors are required to complete the auditing course and apply for accreditation with, for example, RABQSA, the International Registry of Certificated Auditors (IRCA), or RCB.

Table 7.3 summarizes the major similarities and differences between the appraisal methods. In Table 7.4, some of the major requirements of the ISO 20000 certification scheme are compared to those found in the SCAMPI MDD.

Table 7.3 Comparison of ISO and SCAMPI Appraisal Methods

ISO 9001:2000 Certification	*ISO 20000:2005 Certification*	*SCAMPI A*
No specific method description (guidance is provided in ISO 19011)	No specific method description (however, there is an itSMF "certification scheme")	Well-defined, documented method, MDD
Audit scope is defined prior to start of audit and documented in the audit plan; exceptions are noted	Audit scope is defined prior to start of audit and documented in the audit plan; extent of management control is documented	Appraisal scope is defined in the Appraisal Plan
All requirements in the standard are examined	All requirements in the standard are examined	All SPs and GPs for PAs in scope are evaluated
Criteria for judging process implementation are not specified	Criteria for judging process implementation are not specified	Follows strict rules for characterization of practice implementation
Objective evidence collected via documentation reviews, interviews, and observations	Objective evidence collected via documentation reviews and interviews	Objective evidence collected via two types of documents and interviews
Evaluates process documentation, implementation, effectiveness, and its benefit to the organization	Evaluates process documentation, implementation, and effectiveness	Evaluates process implementation and institutionalization
Evidence can be collected from several sources that are in the scope of the audit (no specific projects are selected as the source of evidence)	Evidence can be collected from several sources that are in the scope of the audit, but management control of the processes must be documented (e.g., when outsourcing)	Focus projects are identified; evidence must be collected through project's whole life cycle; at least one project must cover all PAs in the scope of the appraisal
Single authorized individual (registrar) but can use an audit team	Single authorized individual (registrar) but can use an audit team	Appraisal team (4 to 9 members) led by the lead appraiser
Auditors may not act as consultants	Auditors may not act as consultants	Lead appraisers must be external to the appraised organization
Registrars are authorized by the Registrar Accreditation Board (RAB)	Registrars are authorized by the Registered Certification Body (RCB)	Lead appraisers are authorized by the SEI

ISO 9001:2000 Certification	*ISO 20000:2005 Certification*	*SCAMPI A*
Organizations are certified (or registered) upon completion of the audit	Organizations are certified (or registered) upon completion of the audit	Organizations are rated (not "certified") upon completion of the appraisal
Results of the registration and maintained by the registration body	Certificates are issued by RCB	Results of the appraisal are maintained by the SEI

Table 7.4 Comparison of ISO 20000 and SCAMPI Scope Requirements

ISO 20000 Certification Scheme	*SCAMPI A Method Definition Document*
The audit scope must include all processes identified in the standard • The scope must be the whole management system • Boundaries of the assessed organization must make business sense • The scope will be indicated on the certificate	Organizational unit (OU) that will be subject to appraisal must be defined • Projects that will represent the OU must be carefully selected • Projects must cover the whole OU with special emphasis on geographical dispersion • The scope of the appraisal will be indicated on the Appraisal Disclosure Statement (ADS)
Organization must have management control over all processes • Outsourcing is allowable but the organization must specify interfaces with the supplier and must indicate clear responsibilities for those processes	Projects, representing the OU, must follow organizational set of standard processes • There are no specific requirements for appraising suppliers
Organization is not required to control infrastructure, service desk, etc. • Organization must evaluate effects of those processes on its own processes	Control of the suppliers is limited to their products
Audit includes documentation review and interviews • Surveillance audits should be conducted at least once a year; internal audits must be regularly performed • Re-auditing is required every three years	Appraisal examines evidence in documents and interviews; each model practice is characterized, rolled up to the OU, and the rating of goals and process areas are completed; based on these results, a maturity/capability level is assigned • The appraisal result is valid for three years

Conclusions

When considering multiple frameworks and their auditing schemes, the objective is to find sufficient commonality among them to enable collection of objective evidence that would be applicable to all of the schemes. Although there are many similarities among the three audit approaches, there are fundamental differences in the approach to defining the audit scope and selecting the evidence sample.

Let's consider ISO 9001 and CMMI and their audit approaches as shown in Figure 7.4, with evidence collected from sample projects A, B, and C. The SCAMPI A method requires that, at a minimum, one project provides evidence for all PAs across the complete life cycle. Moreover, the SCAMPI A method requires organizations with more than three projects to have at least three documented instances of each project-related practice implementation. The ISO audit has no such requirement. In an ISO audit, evidence showing conformance to an ISO 9001 requirement can be collected from several sources and there is no requirement for substantiating a project life cycle implementation.

What does that imply when developing a common ISO–CMMI evidence collection scheme?

SCAMPI A is more rigorous than ISO audits with more specific evidence collection requirements. Therefore, for an organization that is attempting to achieve a CMMI rating, evidence for an ISO audit will be readily available for those CMMI practices with sufficiently strong mappings to the ISO standard. In contrast, for an ISO-certified organization, evidence collected for certification may

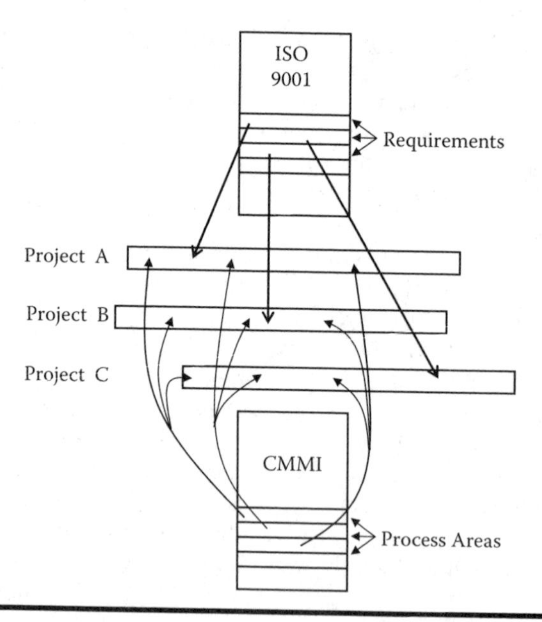

Figure 7.4 Evidence collection in ISO 9001 and CMMI audits.

not be sufficient for the SCAMPI A method. Therefore, the recommended audit approach for an organization seeking ISO 9001 registration and a CMMI rating is to start with CMMI and then extend the evidence to ISO, which is consistent with our CMMI-centric view.

An efficient multiframework appraisal approach can be less burdensome and less expensive than separate appraisals. This can free an organization to stay focused on business goals and spend its time and efforts on true improvements.

Summary

In this chapter we have presented appraisal approaches related to the selected frameworks and described their characteristics. It appears that the SCAMPI A method is more rigorous than the other assessment methods discussed. Therefore, in an organization that is attempting to achieve a CMMI rating, evidence for an ISO audit will be readily available for those CMMI practices with sufficiently strong mappings to the ISO standard. In contrast, for an ISO-certified organization, evidence collected for certification may not be sufficient for the SCAMPI A method. Thus, the recommended audit approach for an organization seeking ISO 9001 registration and a CMMI rating is to start with CMMI and then extend the evidence to ISO. This approach is consistent with our CMMI-centric view. By analyzing framework maps and understanding audit requirements, a detailed audit approach can be developed. As we have repeatedly emphasized, however, the maps are only tools meant to assist in process improvement efforts. They cannot substitute for an investment of time and energy into understanding frameworks and their synergy.

Appendix A: Acronyms

ADS	Appraisal Disclosure Statement
AHP	Analytical Hierarchy Process
ARC	Appraisal Requirements for CMMI
CAR	Causal Analysis and Resolution
CL	Capability Level
CM	Configuration Management
CMM	Capability Maturity Model
CMMI	Capability Maturity Model Integration
CMMI-ACQ	CMMI for Acquisition
CMMI-DEV	CMMI for Development
CMMI-SVC	CMMI for Services
CMU	Carnegie Mellon University
COTS	Commercial Off-the-Shelf
DAR	Decision Analysis and Resolution
EIA	Electronic Industries Alliance
EPG	Engineering Process Group
GG	Generic Goal
GP	Generic Practice
IAF	International Accreditation Forum
IDEAL	Initiating–Diagnosing–Establishing–Acting–Learning
IPD	Integrated Product Development
IPM	Integrated Project Management
IPPD	Integrated Product and Process Development
ISM	Integrated Supplier Management
ISO	International Organization for Standardization
IT	Integrated Teaming
ITIL	Information Technology Infrastructure Library
KPA	Key Process Area
LA	Lead Appraiser
MA	Measurement and Analysis

MAUT	Multi-Attribute Utility Technique
MDD	Method Definition Document
ML	Maturity Level
MSG	Management Steering Group
OEI	Organizational Environment for Integration
OID	Organizational Innovation and Deployment
OPD	Organizational Process Definition
OPF	Organizational Process Focus
OPP	Organizational Process Performance
OT	Organizational Training
OU	Organizational Unit
PA	Process Area
PAL	Process Asset Library
PAT	Process Action Team
PDCA	Plan–Do–Check–Act
PG	Process Group
PI	Product Integration
PMC	Project Monitoring and Control
PP	Project Planning
PPQA	Process and Product Quality Assurance
PRM	Process Reference Model
QPM	Quantitative Project Management
RD	Requirements Development
REQM	Requirements Management
RSKM	Risk Management
SAM	Supplier Agreement Management
SCAMPI	Standard CMMI Appraisal Method for Process Improvement
SECM	System Engineering Capability Model
SEI	Software Engineering Institute
SEPG	Systems Engineering Process Group
SLA	Service Level Agreement
SLM	Service Level Management
SG	Specific Goal
SP	Specific Practice
TS	Technical Solution
TWG	Technical Working Group
VAL	Validation
VER	Verification

Appendix B: References

Ahern 2005

Ahern, Dennis, Jim Armstrong, Aaron Clouse, et al., 2005, *CMMI® SCAMPI Distilled: Appraisals for Process Improvement*, Boston: Addison-Wesley Professional.

Basili 1992

Basili, Victor, 1992, *Software Modeling and Measurement: The Goal/Question/Metric Paradigm*, CS-TR-2956, UMIACS-TR-92-96, University of Maryland.

Boehm 1988

Boehm, Barry, May 1988, A Spiral Model of Software Development and Enhancement, *IEEE Computer*, 21(5), 61–72.

Bush and Dunaway 2005

Bush, Marilyn and Donna Dunaway, 2005, *CMMI® Assessments: Motivating Positive Change*, Boston: Addison-Wesley Professional.

Chrissis et al. 2003

Chrissis, Mary Beth, Gian Wemyss, Dennis Goldenson, et al., 2003, *CMMI Interpretive Guidance Project: Preliminary Report*, CMU/SEI-2003-SR-007, Pittsburgh: Software Engineering Institute, Carnegie Mellon University.

Chrissis et al. 2006

Chrissis, Mary Beth, Mike Konrad, and Sandy Shrum, 2006, *CMMI®: Guidelines for Process Integration and Product Improvement* (2nd ed.), Boston: Addison-Wesley.

CMMI 2006

CMMI Product Team, 2006, *CMMI for Development, Version 1.2*, CMU/SEI-2006-TR-008, Pittsburgh: Software Engineering Institute, Carnegie Mellon University.

Deming 1982

Deming, W. Edwards, 1982, *Out of the Crisis*, Cambridge: Massachusetts Institute of Technology.

Gallagher and Shrum 2004

Gallagher, Brian P. and Sandy Shrum, 2004, Applying CMMI to Systems Acquisition, *Crosstalk*, August 2004, 8–12.

Garcia and Turner 2006

Garcia, Suzanne and Richard Turner, 2006, *CMMI Survival Guide*, Boston: Addison-Wesley.

Hefner 2004

Hefner, Rick, 2004, *Applying CMMI Generic Practices with Good Judgment*, 2004 SEPG Conference, Boston, MA.

Herndon et al. 2003

Herndon, Mary Anne, Robert Moore, Mike Phillips, et al., 2003, *Interpreting Capability Maturity Model Integration (CMMI) for Service Organizations—A Systems Engineering and Integration Services Example*, CMU/SEI-2003-TN-005, Pittsburgh: Software Engineering Institute, Carnegie Mellon University.

ISO 1995

International Organization for Standardization, 1995, *ISO 12207:1995, Information technology — Software life cycle processes*, Geneva.

ISO 1998

International Organization for Standardization, 1998, *ISO/IEC TR 15271:1998, Information technology — Guide for ISO/IEC 12207 (Software Life Cycle Processes).*

ISO 2000a

International Organization for Standardization, 2000, *ISO 9000:2000, Quality management systems — Fundamentals and vocabulary.*

ISO 2000b

International Organization for Standardization, 2000, *ISO 9001:2000, Quality management systems — Requirements.*

ISO 2000c

International Organization for Standardization, 2000, *ISO 9004:2000, Quality management systems — Guidelines for performance improvements.*

ISO 2001a

International Organization for Standardization, 2001, *ISO/TR 10013:2001, Guidelines for quality management system documentation.*

ISO 2001b

International Organization for Standardization, 2001, *ISO/IEC 9126-1:2001, Software engineering — Product quality — Part 1: Quality model.*

ISO 2001c

International Organization for Standardization, 2001, *ISO/IEC 9126-4:2001*

ISO 2002a

International Organization for Standardization, 2002, *ISO 19011:2002, Guidelines for quality and/or environmental management systems auditing.*

ISO 2002b

International Organization for Standardization, 2002, ISO 12207/Amendment 1:2002, *Information technology — Software lifecycle processes.*

ISO 2003
International Organization for Standardization, 2003, *ISO/IEC TR 19760:2003, Systems engineering — A guide for the application of ISO/IEC 15288 (System life cycle processes).*

ISO 2004a
International Organization for Standardization, 2004, *ISO 90003:2004, Software engineering — Guidelines for the application of ISO 9001:2000 to computer software.*

ISO 2004b
International Organization for Standardization, 2004, *ISO/IEC 15504:2004, Information technology — Process assessment* (multiple parts).

ISO 2005a
International Organization for Standardization, 2005, *ISO/IEC 20000-1, Information technology — Service management — Part 1: Specification.*

ISO 2005b
International Organization for Standardization, 2005, *ISO/IEC 20000-2, Information technology — Service management — Part 2: Code of practice.*

ISO 2005c
International Organization for Standardization, 2005, *ISO/IEC 17799:2005, Information technology — Security techniques — Code of practice for information security management.*

ISO 2007
International Organization for Standardization, 2007, *ISO/IEC 15939:2007, Systems and software engineering — Measurement process.*

ISO 2008a
International Organization for Standardization, 2008, *ISO/IEC 15288:2008, Systems and software engineering — System life cycle processes.*

ISO 2008b
International Organization for Standardization, 2008, *ISO/IEC 12207:2008, Systems and software engineering — Software life cycle processes.*

ISO 2008c
International Organization for Standardization, 2008 (target date), *ISO/IEC 24748, Systems and Software engineering — Life cycle management — Guide for life cycle management.*

Kasse 2002
Kasse, Tim, 2002, *Constagedeous Approach to Process Improvement*, 2002 SEPG Conference, Phoenix, AZ.

Kasse 2004
Kasse, Tim, 2004, *Practical Insight into CMMI*, Boston: Artech House.

Kitson et al. 2007

Kitson, David H., Robert Vickroy, John Walz, and Dave Wynn, 2007, *An Initial Comparative Analysis of the CMMI Product Suite (v1.2) and ISO 9000 Family*, SEI Special Report, November 2007, draft, Pittsburgh: Software Engineering Institute, Carnegie Mellon University.

Kulpa and Johnson 2003

Kulpa, Margaret and Kent Johnson, 2003, *Interpreting the CMMI*, Boca Raton, FL: Auerbach Publications.

McFeeley 1996

McFeeley, Robert, 1996, *IDEAL: A User's Guide for Software Process Improvement*, CMU/SEI-96-HB-001, Pittsburgh: Software Engineering Institute, Carnegie Mellon University.

Mutafelija and Stromberg 2003a

Mutafelija, Boris and Harvey Stromberg, 2003, *Systematic Process Improvement Using ISO 9001:2000 and CMMI*, Boston: Artech House.

Mutafelija and Stromberg 2003b

Mutafelija, Boris and Harvey Stromberg, 2003, *Exploring CMMI-ISO 9001:2000 Synergy When Developing a Process Improvement Strategy*, SEPG 2003, Boston, MA.

SCAMPI 2006a

SCAMPI A Upgrade Team, 2006, *Appraisal Requirements for CMMI®, Version 1.2 (ARC, V1.2)*, CMU/SEI-2006-TR-011, Pittsburgh: Software Engineering Institute, Carnegie Mellon University.

SCAMPI 2006b

SCAMPI A Upgrade Team, 2006, *Standard CMMI® Appraisal Method for Process Improvement (SCAMPI℠) A, Version 1.2: Method Definition Document*, CMU/SEI-2006-HB-002, Pittsburgh: Software Engineering Institute, Carnegie Mellon University.

Schlickman 2003

Schlickman, Jay, 2003, *ISO 9001:2000 Quality Management System Design*, Boston: Artech House.

Siviy et al. 2007

Siviy, Jeannine M., M. Lynn Penn, and Robert Stoddard, 2007, *Achieving Success via Multi-Model Process Improvement*, SEPG 2007, Austin, TX.

Stromberg and Mutafelija 2002

Stromberg, Harvey and Boris Mutafelija, 2002, *Using the CMMI When Implementing ISO 9001:2000 for Software*, SEPG 2002, Phoenix, AZ.

Stromberg and Mutafelija 2004

Stromberg, Harvey and Boris Mutafelija, 2004, *Preserving the CMM Investment When Transitioning to CMMI*, SEPG 2004, Orlando, FL.

Taylor 1911

Taylor, Frederick W., 1911, *The Principles of Scientific Management*, New York: Harper & Brothers.

Appendix C: Changes from CMMI v1.1 to CMMI v1.2

Note: Advanced practices were eliminated in CMMI v1.2.

PA	*CMMI v1.1*	*CMMI v1.2*
IPM	**SG 1** The project is conducted using a defined process that is tailored from the organization's set of standard processes.	
	SP 1.1-1 Establish and maintain the project's defined process.	SP 1.1 Establish and maintain the project's defined process form project startup through the life of the project. (Project startup emphasis added)
	SP 1.2-1 Use the organizational process assets and measurement repository for estimating and planning the project's activities.	No change
		SP 1.3 Establish and maintain the project's work environment based on the organization's work environment standards. (New SP)

PA	*CMMI v1.1*	*CMMI v1.2*
	SP 1.3-1 Integrate the project plan and the other plans that affect the project to describe the project's defined process.	Numbering changed to SP 1.4
	SP 1.4-1 Manage the project using the project plan, the other plans that affect the project, and the project's defined process.	Numbering changed to SP 1.5
	SP 1.5-1 Contribute work products, measures, and documented experiences to the organizational process assets.	Numbering changed to SP 1.6; subpractice 5 added
IPM	**SG 2** Coordination and collaboration of the project with relevant stakeholders is conducted.	
	SP 2.1-1 Manage the involvement of the relevant stakeholders in the project.	Small changes
	SP 2.2-1 Participate with relevant stakeholders to identify, negotiate, and track critical dependencies.	No change
	SP 2.3-1 Resolve issues with relevant stakeholders.	No change
IPM	**SG 3** The project is conducted using the project's shared vision.	SG 3 Apply IPPD Principles. (New goal—IPPD addition)
	SP 3.1-1 Identify expectations, constraints, interfaces, and operational conditions applicable to the project's shared vision.	SP 3.1 Establish and maintain a shared vision for the project. (SP rewritten using some material from IPM SP 3.1-1 and 3.2-1)
	SP 3.2-1 Establish and maintain a shared vision for the project.	
		SP 3.2 Establish and maintain the integrated team structure for the project. (Moved from SP 4.1-1 and rewritten)
		SP 3.3 Allocate requirements, responsibilities, tasks, and interfaces to teams in the integrated team structure. (Moved from SP 4.2-1 and rewritten)

PA	*CMMI v1.1*	*CMMI v1.2*
		SP 3.4 Establish and maintain integrated teams in the structure. (Moved from SP 4.3-1 and rewritten using some material from IT SP 2.2-1) SP 3.5 Ensure collaboration among interfacing teams. (Moved from IT SP 2.5-1 and rewritten)
IPM	**SG 4** Organize Integrated Teams for IPPD.	Goal eliminated
	SP 4.1-1 Determine the integrated team structure that will best meet the project objectives and constraints.	Moved to SP 3.2
	SP 4.2-1 Develop a preliminary distribution of requirements, responsibilities, authorities, tasks, and interfaces to teams in the selected integrated team structure.	Moved to SP 3.3
	SP 4.3-1 Establish and maintain teams in the integrated team structure.	Moved to SP 3.4
ISM	**SG 1** Potential sources of products that best fit the needs of the project are identified, analyzed, and selected.	Goal eliminated
	SP 1.1-1 Identify and analyze potential sources of products that may be used to satisfy the project's requirements.	Deleted
	SP 1.2-1 Use a formal evaluation process to determine which sources of custom-made and off-the-shelf products to use.	Deleted
ISM	**SG 2** Work is coordinated with suppliers to ensure the supplier agreement is executed appropriately.	Goal eliminated
	SP 2.1-1 Monitor and analyze selected processes used by the supplier.	Moved to SAM SP 2.2
	SP 2.2-1 For custom-made products, evaluate selected supplier work products.	Moved to SAM SP 2.3
	SP 2.3-1 Revise the supplier agreement or relationship, as appropriate, to reflect changes in conditions.	Deleted

PA	*CMMI v1.1*	*CMMI v1.2*
OPD	**SG 1** A set of organizational process assets is established and maintained.	
	SP 1.1-1 Establish and maintain the organization's set of standard processes.	No change
	SP 1.2-1 Establish and maintain descriptions of the life-cycle models approved for use in the organization.	No change
	SP 1.3-1 Establish and maintain the tailoring criteria and guidelines for the organization's set of standard processes.	No change
	SP 1.4-1 Establish and maintain the organization's measurement repository.	No change
	SP 1.5-1 Establish and maintain the organization's process asset library.	No change
	SP 1.6 Establish and maintain work environment standards.	New
OPD	**SG 2** Organizational rules and guidelines that govern the operation of integrated teams are provided.	New goal—IPPD addition
	SP 2.1 Establish and maintain empowerment mechanisms to enable timely decision making.	Moved from OEI and rewritten
	SP 2.2 Establish and maintain organizational rules and guidelines for structuring and forming integrated teams.	Moved from OEI and rewritten
	SP 2.3 Establish and maintain organizational guidelines to help team members balance their team and home organization responsibilities.	Moved from OEI and rewritten
OPF	**SG 1** Strengths, weaknesses, and improvement opportunities for the organization's processes are identified periodically and as needed.	
	SP 1.1-1 Establish and maintain the description of the process needs and objectives for the organization.	No change
	SP 1.2-1 Appraise the processes of the organization periodically and as needed to maintain an understanding of their strengths and weaknesses.	Changed wording to: organization's processes

PA	*CMMI v1.1*	*CMMI v1.2*
	SP 1.3-1 Identify improvements to the organization's processes and process assets.	No change
OPF	**SG 2** Process actions that address improvements to the organization's processes and process assets are planned and implemented.	Slight wording change
	SP 2.1-1 Establish and maintain process action plans to address improvements to the organization's processes and process assets.	No change
	SP 2.2-1 Implement process action plans across the organization.	No change
	SP 2.3-1 Deploy organizational process assets across the organization.	Moved to SP 3.1
	SP 2.4-1 Incorporate process-related work products, measures, and improvement information derived from planning and performing the process into the organizational process assets.	Moved to SP 3.4
OPF	**SG 3** The organizational process assets are deployed across the organization and process-related experiences are incorporated into the organizational process assets.	New goal
	SP 3.1 Deploy organizational process assets across the organization.	Moved from SP 2.3-1
	SP 3.2 Deploy the organization's set of standard processes to projects at their startup and deploy changes to them as appropriate throughout the life of each project.	New
	SP 3.3 Monitor the implementation of the organization's set of standard processes and use of process assets on all projects.	New
	SP 3.4 Incorporate process-related work products, measures, and improvement information derived from planning and performing the process into the organizational process assets.	Moved from SP 2.4-1

PA	*CMMI v1.1*	*CMMI v1.2*
RD	**SG 1** Stakeholder needs, expectations, constraints, and interfaces are collected and translated into customer requirements.	
	SP 1.1-1 Identify and collect stakeholder needs, expectations, constraints, and interfaces for all phases of the product life cycle.	SP combined as SP 1.1
	SP 1.1-2 Elicit stakeholder needs, expectations, constraints, and interfaces for all phases of the product's life cycle.	SPs combined as SP 1.1
	SP 1.2-1 Transform stakeholder needs, expectations, constraints, and interfaces into customer requirements.	No change
RD	**SG 2** Customer requirements are refined and elaborated to develop product and product component requirements for the product life cycle.	
	SP 2.1-1 Establish and maintain product and product-component requirements which are based on the customer requirements.	No change
	SP 2.2-1 Allocate the requirements for each product component.	No change
	SP 2.3-1 Identify interface requirements.	No change
RD	**SG 3** The requirements are analyzed and validated, and a definition of required functionality is developed.	
	SP 3.1-1 Establish and maintain operational concepts and associated scenarios.	No change
	SP 3.2-1 Establish and maintain a definition of required functionality.	No change
	SP 3.3-1 Analyze requirements to ensure that they are necessary and sufficient.	No change
	SP 3.4-3 Analyze requirements to balance stakeholder needs and constraints.	No change
	SP 3.5-1 Validate requirements to ensure the resulting product will perform appropriately in its intended environment.	SPs combined as SP 3.5

PA	CMMI v1.1	CMMI v1.2
	SP 3.5-2 Validate requirements to ensure the resulting product will perform as intended in the user's environment using multiple techniques as appropriate.	SPs combined as SP 3.5
REQM	**SG 1** Requirements are managed and inconsistencies with project plans and work products are identified.	
	SP 1.1-1 Develop an understanding with the requirements providers on the meaning of the requirements.	No change
	SP 1.2-2 Obtain commitment to the requirements from the project participants.	No change
	SP 1.3-1 Manage changes to the requirements as they evolve during the project.	No change
	SP 1.4-2 Maintain bi-directional traceability among the requirements and the project plans and work products.	SP 1.4 Maintain bidirectional traceability among the requirements and work products. (Wording changed)
	SP 1.5-1 Identify inconsistencies between the project plans and work products and the requirements.	No change
SAM	**SG 1** Agreements with the suppliers are established and maintained.	
	SP 1.1-1 Determine the type of acquisition for each product or product component to be acquired.	No change
	SP 1.2-1 Select suppliers based on an evaluation of their ability to meet the specified requirements and established criteria.	No change
	SP 1.3-1 Establish and maintain formal agreements with the supplier.	No change
SAM	**SG 2** Agreements with the suppliers are satisfied by both the project and the supplier.	
	SP 2.1-1 Review candidate COTS products to ensure that they satisfy the specified requirements that are covered under a supplier agreement.	Moved to TS SP 1.1, subpractice 1

PA	CMMI v1.1	CMMI v1.2
	SP 2.2-1 Perform activities with the supplier as specified in the supplier agreement.	Numbering changed to SP 2.1
		SP 2.2 Select, monitor, and analyze processes used by the supplier. (Moved from ISM, SP 2.1)
		SP 2.3 Select and evaluate work products from the supplier of custom-made products. (Moved from ISM, SP 2.2)
	SP 2.3-1 Ensure that the supplier agreement is satisfied before accepting the acquired product.	Numbering changed to SP 2.4
	SP 2.4-1 Transition the acquired products from the supplier to the project.	Numbering changed to SP 2.5
TS	**SG 1** Product or product component solutions, including applicable product-related processes, are selected from alternative solutions.	
	SP 1.1-1 Develop alternative solutions and selection criteria.	SPs combined as SP 1.1
	SP 1.1-2 Develop detailed alternative solutions and selection criteria.	SPs combined as SP 1.1 Develop alternative solutions and selection criteria.
	SP 1.2-2 Evolve the operational concept, scenarios, and environments to describe the conditions, operating modes, and operating states specific to each product component.	Moved to RD SP 2.1
	SP 1.3-1 Select the product component solutions that best satisfy the criteria established.	SP numbering changed to SP 1.2
TS	**SG 2** Product or product component designs are developed.	
	SP 2.1-1 Develop a design for the product or product component.	No change
	SP 2.2-3 Establish and maintain a technical data package.	No change
	SP 2.3-1 Establish and maintain the solution for product-component interfaces.	SPs combined as SP 2.3

PA	*CMMI v1.1*	*CMMI v1.2*
	SP 2.3-3 Design comprehensive product component interfaces in terms of established and maintained criteria.	SPs combined as SP 2.3 Design product component interfaces using established criteria.
	SP 2.4-3 Evaluate whether the product components should be developed, purchased, or reused based on established criteria.	Advanced practice maintained
TS	**SG 3** Product components, and associated support documentation, are implemented from their designs.	
	SP 3.1-1 Implement the designs of the product components.	No change
	SP 3.2-1 Develop and maintain the end-use documentation.	No change
VAL	**SG 1** Preparation for validation is conducted.	
	SP 1.1-1 Select products and product components to be validated and the validation methods that will be used for each.	No change
	SP 1.2-2 Establish and maintain the environment needed to support validation.	No change
	SP 1.3-3 Establish and maintain procedures and criteria for validation.	No change
VAL	**SG 2** The product or product components are validated to ensure that they are suitable for use in their intended operating environment.	
	SP 2.1-1 Perform validation on the selected products and product components.	No change
	SP 2.2-1 Analyze the results of the validation activities and identify issues.	"and identify issues"—deleted
VER	**SG 1** Preparation for verification is conducted.	
	SP 1.1-1 Select the work products to be verified and the verification methods that will be used for each.	No change
	SP 1.2-2 Establish and maintain the environment needed to support verification.	No change

PA	*CMMI v1.1*	*CMMI v1.2*
	SP 1.3-1 Establish and maintain verification procedures and criteria for the selected work products.	No change
VER	**SG 2** Peer reviews are performed on selected work products.	
	SP 2.1-1 Prepare for peer reviews of selected work products.	No change
	SP 2.2-3 Conduct peer reviews on selected work products and identify issues resulting from the peer review.	No change
	SP 2.3-1 Analyze data about preparation, conduct, and results of the peer reviews.	No change
VER	**SG 3** Selected work products are verified against their specific requirements.	
	SP 3.1-1 Perform verification on the selected work products.	No change
	SP 3.2-3 Analyze the results of all verification activities and identify corrective actions.	"and identify corrective actions"—deleted from the title and statement of this practice

Appendix D: ISO 9001:2000 to CMMI v1.2 Map

Table D.1 ISO 9001:2000 Section 4, Quality Management System

Section	*ISO 9001:2000 Requirement Keywords*	*CMMI PAs*	*CMMI Practices*	*Confidence*	*Comment*
4	**Quality management system**				
4.1	**General requirements**				
	Establish QMS	All PAs	GP 2.1	100	
		OPD	SP 1.1	100	
		OPD	SP 1.2	100	
		OPD	SP 1.3	100	
		OPF	SP 3.1	100	
		OPF	SP 3.2	100	
		OPF	SP 3.3	100	
		OPF	SP 3.4	100	
	Identify processes	OPD	SP 1.1	100	
	Determine sequence	OPD	SP 1.1	100	
	Effective operation	OID	SP 2.1	30	
		OPF	SP 3.3	30	
	Resources	All PAs	GP 2.3	100	
	Monitor processes	All PAs	GP 2.8	100	
		All PAs	GP 2.9	100	
	Implement actions	OPF	SP 2.1	100	
		OPF	SP 2.2	100	
	Manage using ISO standard	All PAs	GP 2.1	60	Organizational policies must address ISO 9001
	Control outsourced processes	SAM	SP 1.3	100	
		SAM	SP 2.1	100	

		SAM	SP 2.2	100	
	Outsourced process control in QMS	SAM	GP 2.2	100	
		SAM	SP 1.3	100	
		SAM	SP 2.2	100	
		SAM	SP 2.3	100	
4.2	**Documentation requirements**				
4.2.1	**General**				
	Document quality policy	All PAs	GP 2.1	100	
	Document quality manual	OPD	SP 1.1	100	
		OPD	SP 1.2	100	
		OPD	SP 1.3	100	
	Document procedures	OPD	SP 1.1	100	
		OPD	SP 1.2	100	
		OPD	SP 1.3	100	
	Records	PP	SP 2.3	30	
		PPQA	SP 2.2	100	
4.2.2	**Quality Manual**				
	Establish Quality Manual				
	QMS scope	OPD	SP 1.1	60	
	Establish QMS procedures	All PAs	GP 2.2	60	
	Describe process interaction	OPD	SP 1.1	100	See subpractice 3
4.2.3	**Control of documents**				
	Control required documents	All PAs	GP 2.6	100	
		PMC	SP 1.4	100	
		PP	SP 2.3	100	

(continued)

Table D.1 ISO 9001:2000 Section 4, Quality Management System (continued)

Section	*ISO 9001:2000 Requirement Keywords*	*CMMI PAs*	*CMMI Practices*	*Confidence*	*Comment*
	Control records	PMC	SP 1.4	60	
		PP	SP 2.3	60	
	Document control procedure	CM	GP 2.2	60	
	Approve documents	CM	GP 2.2	60	
	Review and update	CM	SP 2.2	30	
	Identify changes	CM	SP 2.2	100	
	Relevant versions available	All PAs	GP 2.6	100	
	Identifiable documents			0	
	Control external documents	CM	SP 1.1	100	
	Obsolete documents	CM	SP 3.2	30	
	Identify documents	CM	SP 1.1	100	
4.2.4	**Control of records**				
	Records provide evidence of conformity	PPQA	SP 2.2	100	ISO 9001 requirement partially addressed. Conformity is generally addressed in SCAMPI.
		PP	SP 2.3	60	
	Records identifiable			0	
	Record control procedure	All PAs	GP 2.6	60	
		CM	GP 2.2	100	

Table D.2 ISO 9001:2000 Section 5, Management Responsibility

Section	*ISO 9001:2000 Requirement Keywords*	*CMMI PAs*	*CMMI Practices*	*Confidence*	*Comment*
5	**Management responsibility**				
5.1	**Management commitment**				
	Communicate importance	All PAs	GP 2.1	60	
	Quality policy	All PAs	GP 2.1	60	
	Quality objectives	All PAs	GP 2.10	100	This addresses evidence of commitment. Specific objectives are addressed in section 5.4.1.
	Management review	All PAs	GP 2.10	100	
	Resource availability	All PAs	GP 2.3	100	
5.2	**Customer focus**				
	Determine customer requirements	RD	GP 2.1	100	CMMI does not focus on enhancing customer satisfaction.
		RD	GP 2.10	100	
		REQM	GP 2.1	100	
		REQM	GP 2.10	100	
5.3	**Quality policy**				
	Top management quality policy responsibility				
	Appropriate to organization	All PAs	GP 2.1	100	
		OPF	SP 1.1	100	
	Commitment to comply	OPF	GP 2.1	100	
	Framework for quality objectives	MA	GP 2.10	100	
		OPF	GP 2.1	100	
		OPF	GP 2.10	100	

(continued)

Table D.2 ISO 9001:2000 Section 5, Management Responsibility (continued)

Section	*ISO 9001:2000 Requirement Keywords*	*CMMI PAs*	*CMMI Practices*	*Confidence*	*Comment*
		PPQA	GP 2.10	100	
	Communicated	OPF	GP 2.1	100	
5.4	**Planning**				
5.4.1	**Quality objectives**				
	Quality objectives established	MA	GP 2.10	100	
		MA	SP 1.1	100	
		OPF	GP 2.10	100	
		OPF	SP 1.1	100	
		OPP	GP 2.10	100	
		OPP	SP 1.3	100	
		QPM	GP 2.10	100	
		QPM	SP 1.1	100	
	Measurable objectives	OPP	SP 1.3	100	
		QPM	SP 1.1	100	
5.4.2	**Quality management system planning**				
	Plan to meet quality objectives	OPD	GP 2.10	100	
		OPF	GP 2.10	100	
	Maintain QMS integrity	OPD	GP 2.6	60	
5.5	**Responsibility, authority and communication**				
5.5.1	**Responsibility and authority**				
	Top management defines responsibility	All PAs	GP 2.4	100	

5.5.2	**Management representative**			
	Appoint member of management	OPF	GP 2.4	100
	Establish QMS processes	OPD	GP 2.4	100
		OPF	GP 2.4	100
	Report performance of QMS	OPF	GP 2.10	100
	Customer requirement awareness	RD	GP 2.10	60
		REQM	GP 2.10	60
5.5.3	**Internal communication**			
	Establish communication processes	OPD	GP 2.10	100
		OPF	GP 2.1	100
		OPF	GP 2.10	100
5.6	**Management review**			
5.6.1	**General**			
	Review QMS	OPD	GP 2.10	100
		OPF	GP 2.10	100
	Assess improvement opportunities	OPF	GP 2.10	100
		OPF	SP 1.2	100
		OPF	SP 1.3	100
	Maintain records	OPD	GP 2.6	60
		OPF	GP 2.6	60
5.6.2	**Review input**			
	Audit	PPQA	GP 2.10	100
	Customer			0
	Conformity	PPQA	GP 2.10	100
	Preventive action	CAR	GP 2.10	60
	Follow-up	All PAs	GP 2.10	100

(continued)

Table D.2 ISO 9001:2000 Section 5, Management Responsibility (continued)

Section	*ISO 9001:2000 Requirement Keywords*	*CMMI PAs*	*CMMI Practices*	*Confidence*	*Comment*
	QMS plans	OPF	GP 2.10	100	
	Improvement	OID	GP 2.10	100	
		OPF	GP 2.10	100	
5.6.3	**Review output**				
	Improve effectiveness	OPF	GP 2.10	60	
	Improve product	All PAs	GP 2.10	60	
	Resources	All PAs	GP 2.3	100	
		All PAs	GP 2.10	100	

Table D.3 ISO 9001:2000 Section 6, Resource Management

Section	*ISO 9001:2000 Requirement Keywords*	*CMMI PAs*	*CMMI Practices*	*Confidence*	*Comment*
6	**Resource Management**				
6.1	**Provision of resources**				
	Implement and maintain QMS	All PAs	GP 2.3	100	
		OPD	SP 1.6	60	
	Enhance customer satisfaction	All PAs	GP 2.3	30	Customer satisfaction is not explicitly addressed but is implied through many PAs.
6.2	**Human resources**				
6.2.1	**General**				
	Staff has needed skills	All PAs	GP 2.5	100	
		OT	All SPs	100	
6.2.2	**Competence, awareness and training**				
	Determine competence	OT	SP 1.1	100	
		OT	SP 1.2	30	
		PP	SP 2.5	100	
	Provide training	All PAs	GP 2.5	100	
		OT	SP 1.3	100	
		OT	SP 1.4	100	
		OT	SP 2.1	100	
	Evaluate effectiveness	OT	SP 2.3	100	
	Ensure awareness of importance	IPM	SP 3.1	100	Uses IPPD
	Maintain records	OT	SP 2.2	100	

(continued)

Table D.3 ISO 9001:2000 Section 6, Resource Management (continued)

Section	*ISO 9001:2000 Requirement Keywords*	*CMMI PAs*	*CMMI Practices*	*Confidence*	*Comment*
6.3	**Infrastructure**				
	Provide services and equipment	All PAs	GP 2.3	100	
		IPM	SP 1.3	100	
		OPD	SP 1.6	100	
		PI	SP 1.2	100	
		PP	SP 2.4	100	
		VAL	SP 1.2	100	
		VER	SP 1.2	100	
6.4	**Work environment**				
	Maintain environment to meet requirements	All PAs	GP 2.3	100	
		IPM	SP 1.3	100	
		OPD	SP 1.6	100	
		PI	SP 1.2	100	
		PP	SP 2.4	100	
		VAL	SP 1.2	100	
		VER	SP 1.2	100	

Table D.4 ISO 9001:2000 Section 7, Product Realization

Section	*ISO 9001:2000 Requirement Keywords*	*CMMI PAs*	*CMMI Practices*	*Confidence*	*Comment*
7	**Product realization**				
7.1	**Planning product realization**				
	Develop needed processes	OPD	SP 1.1	100	
		OPD	SP 1.2	100	
		OPD	SP 1.3	100	
	Planning is consistent with other processes	All PAs	GP 2.2		
		All PAs	GP 3.1	100	
		IPM	SP 1.1	100	
		IPM	SP 1.4	100	
		OPD	SP 1.1	100	
		OPD	SP 1.2	100	
		OPD	SP 1.3	100	
		PP	SP 2.7	100	
		PP	SP 3.1	100	
	Quality objectives	RD	SP 1.2	100	
		RD	SP 2.1	100	
		RD	SP 2.3	100	
		RD	SP 3.2	100	
		QPM	SP 1.1	100	
	Establish processes	IPM	SP 1.1	100	
		PP	GP 2.2	100	

(continued)

Table D.4 ISO 9001:2000 Section 7, Product Realization (continued)

Section	*ISO 9001:2000 Requirement Keywords*	*CMMI PAs*	*CMMI Practices*	*Confidence*	*Comment*
	Determine activities	IPM	SP 1.1	100	
		PP	SP 1.3	100	
	Determine records	All PAs	GP 2.6	100	
		PP	SP 2.3	100	
	Plans in appropriate format	All PAs	GP 2.2	100	
		PP	SP 2.7	100	
7.2	**Customer-related processes**				
7.2.1	**Determination of requirements related to the product**				
	Customer requirements	RD	SP 1.1	100	
	Unstated requirements	RD	SP 1.2	100	
		RD	SP 2.1	100	
		RD	SP 2.3	100	
		RD	SP 3.1	100	
		RD	SP 3.2	100	
	Statutory requirements	RD	SP 1.1	100	
	Additional requirements			0	
7.2.2	**Review of requirements related to the product**				
	Organization reviews requirements	RD	SP 3.3	100	
		RD	SP 3.4	100	
		RD	SP 3.5	100	
		RD	GP 2.7	100	

		REQM	SP 1.1	100	
		REQM	GP 2.7	100	
	Review before commitment	REQM	SP 1.2	100	
	Requirements are defined	RD	SP 3.3	100	
		REQM	SP 1.1	100	
	Differing requirements	REQM	SP 1.3	30	
	Able to meet requirements	RD	SP 3.3	30	
	Records are kept	RD	GP 2.6	100	
		REQM	GP 2.6	100	
	Confirm understanding of requirements	RD	SP 1.2	100	
		RD	SP 2.1	30	
		RD	SP 3.5	100	
		REQM	SP 1.1	100	
	Keep documents current when requirements change	REQM	SP 1.3	100	
		REQM	SP 1.5	100	
7.2.3	**Customer communication**				
	Communicate product information	RD	GP 2.7	100	
		RD	SP 1.1	100	
		RD	SP 1.2	100	
		RD	SP 3.1	100	
		RD	SP 3.5	100	
		REQM	SP 1.1	100	
	Inquiries	IPM	SP 2.1	30	
	Customer feedback	All PAs	GP 2.7	30	Customer complaints not explicitly addressed in CMMI.

(continued)

Table D.4 ISO 9001:2000 Section 7, Product Realization (continued)

Section	*ISO 9001:2000 Requirement Keywords*	*CMMI PAs*	*CMMI Practices*	*Confidence*	*Comment*
7.3	**Design and development**				
7.3.1	**Design and development planning**				
	Plan design and development	IPM	SP 1.4	100	This mapping is guided by material in ISO 9004 and ISO 90003.
		IPM	SP 1.5	100	
		PI	SP 1.1	100	
		PP	SP 2.7	100	
		TS	GP 2.2	100	
		VAL	SP 1.1	100	
		VER	SP 1.1	100	
	Determine stages	IPM	SP 1.1	100	
		PI	GP 2.2	100	
		PP	SP 1.3	100	
		TS	GP 2.2	100	
	Determine verification and validation	VAL	SP 1.1	100	
		VER	SP 1.1	100	
	Determine responsibility	PI	GP 2.4	100	
		TS	GP 2.4	100	
	Manage interfaces	IPM	SP 2.1	100	
		IPM	SP 2.2	100	
		IPM	SP 2.3	100	
		IPM	SP 3.5	100	
		IPM	GP 2.7	100	
		PI	GP 2.7	100	

		RD	GP 2.7	100
		TS	GP 2.7	100
		VAL	GP 2.7	100
		VER	GP 2.7	100
	Update plans during development	IPM	SP 1.1	100
		IPM	SP 1.4	100
		PP	All SPs	100
7.3.2	**Design and development inputs**			
	Determine inputs to development processes	RD	SP 1.1	100
		RD	SP 1.2	100
		RD	SP 2.1	100
		RD	SP 3.2	100
	Functional requirements	RD	SP 2.1	100
		RD	SP 3.2	100
	Statutory requirements	RD	SP 1.1	100
		RD	SP 1.2	100
	Similar designs	IPM	SP 1.2	60
	Other requirements	RD	SP 1.2	100
		RD	SP 2.1	100
		RD	SP 2.3	100
		RD	SP 3.1	100
	Review inputs	RD	SP 3.3	100
		RD	SP 3.4	60
		RD	SP 3.5	100

(continued)

Table D.4 ISO 9001:2000 Section 7, Product Realization (continued)

Section	*ISO 9001:2000 Requirement Keywords*	*CMMI PAs*	*CMMI Practices*	*Confidence*	*Comment*
	Requirements are consistent and clear	RD	SP 3.3	100	
		RD	SP 3.4	100	
		RD	SP 3.5	100	
7.3.3	**Design and development outputs**				
	Outputs are verifiable	TS	SP 2.1	100	
		TS	SP 2.2	100	
		TS	SP 2.3	100	
		TS	SP 3.1	100	
		TS	SP 3.2	100	
	Outputs approved	TS	SP 3.1	30	
		TS	SP 3.2	30	
	Meet input requirements	TS	SP 1.1	100	
		TS	SP 1.2	100	
		TS	SP 2.1	100	
		TS	SP 3.1	100	
		TS	SP 3.2	100	
	Provide information	TS	SP 1.1	30	
		TS	SP 1.2	30	
		TS	SP 2.2	100	
		TS	SP 2.4	100	
	Acceptance criteria	VAL	SP 1.3	60	
		VER	SP 1.3	60	
	Specify characteristics	TS	SP 2.2	60	

7.3.4	**Design and development review**			
	Development reviewed and evaluated	PMC	SP 1.6	100
		PMC	SP 1.7	100
		PMC	SP 2.1	100
	Identify problems	PMC	SP 1.2	60
		PMC	SP 1.6	100
		PMC	SP 1.7	100
		PMC	SP 2.1	100
	Appropriate functions participate in reviews	IPM	SP 2.1	100
		IPM	SP 2.2	30
		IPM	SP 2.3	30
		PMC	GP 2.7	100
	Records of review are kept	IPM	GP 2.6	60
		PMC	GP 2.6	100
7.3.5	**Design and development verification**			
	Ensure requirements are met	VER	All SPs	100
	Keep verification records	VER	GP 2.6	100
7.3.6	**Design and development validation**			
	Validation follows plans	VAL	All SPs	100
	Validate before delivery	RD	SP 3.5	100
	Keep validation records	RD	GP 2.6	100
		VAL	GP 2.6	100
7.3.7	**Control of design and development changes**			
	Identify changes	CM	SP 3.1	100
		PI	GP 2.6	60

(continued)

Table D.4 ISO 9001:2000 Section 7, Product Realization (continued)

Section	*ISO 9001:2000 Requirement Keywords*	*CMMI PAs*	*CMMI Practices*	*Confidence*	*Comment*
		TS	GP 2.6	60	
	Review and approve changes	CM	SP 2.1	100	
		CM	SP 2.2	100	
	Evaluate effect of changes	CM	SP 2.1	100	
		CM	SP 2.2	100	
	Keep records of changes	CM	SP 3.1	100	
		PI	GP 2.6	60	
		TS	GP 2.6	60	
7.4	**Purchasing**				
7.4.1	**Purchasing process**				
	Purchased product meets requirements	PI	SP 3.1	100	
		SAM	SP 2.4	100	
	Control of supplier depends on product	SAM	SP 2.1	100	
		SAM	SP 2.2	100	
		SAM	SP 2.3	100	
		TS	SP 2.4	30	
	Suppliers selected based on ability	SAM	SP 1.2	100	
	Selection criteria established	SAM	SP 1.2	100	
			SP 1.3	100	
	Records of evaluations kept	SAM	GP 2.6	100	
7.4.2	**Purchasing information**				
	Product requirements described	SAM	SP 1.3	100	
		TS	SP 1.1	100	

	Approval requirements	SAM	SP 1.3	100
	Personnel	SAM	SP 1.3	60
	QMS	SAM	SP 1.3	60
	Adequate requirements described	SAM	SP 1.3	100
7.4.3	**Verification of purchased product**			
	Ensure product meets requirements	SAM	SP 1.3	100
		SAM	SP 2.1	100
		SAM	SP 2.2	100
		SAM	SP 2.3	100
		SAM	SP 2.4	100
		VER	SP 3.1	60
	Supplier site verification	SAM	SP 1.3	60
7.5	**Production and service provision**			
7.5.1	**Control of production and service provision**			
	Plan and implement service provision	PI	GP 2.2	100
		PI	SP 3.1	100
		PI	SP 3.2	100
		PI	SP 3.3	100
		PI	SP 3.4	100
		TS	GP 2.2	100
		TS	SP 3.1	100
		TS	SP 3.2	100
	Product characteristics	PI	SP 2.1	100
		TS	SP 2.2	100
	Work instructions	PI	SP 1.3	100
		TS	GP 2.2	100

(continued)

Table D.4 ISO 9001:2000 Section 7, Product Realization (continued)

Section	*ISO 9001:2000 Requirement Keywords*	*CMMI PAs*	*CMMI Practices*	*Confidence*	*Comment*
	Equipment	PI	GP 2.3	100	
		TS	GP 2.3	100	
	Availability of monitoring devices	PI	GP 2.3	100	
		PI	GP 2.8	100	
		TS	GP 2.3	100	
		TS	GP 2.8	100	
	Implementation of monitoring	PI	GP 2.8	100	
		TS	GP 2.8	100	
	Release activities	CM	SP 1.3	60	CMMI doesn't address postdelivery activities.
		PI	SP 3.4	60	
7.5.2	**Validation of processes for production and service provision**				
	Validate production and service processes	VAL	All SPs	100	
	Demonstrate ability to meet planned results	VAL	All SPs	100	
	Establish review criteria	VAL	SP 1.3	30	
	Approve equipment	VAL	SP 1.2	100	
	Use specific methods	VAL	SP 1.3	100	
	Records	VAL	GP 2.6	100	
		VAL	SP 2.2	100	
	Revalidation	VAL	SP 2.1	100	

7.5.3	**Identification and traceability**			
	Identify products	CM	SP 1.1	100
		CM	SP 2.1	60
		CM	SP 2.2	100
	Identify product status	MA	SP 2.4	100
		PI	GP 2.8	100
		PI	SP 3.3	100
		TS	GP 2.8	100
		VAL	GP 2.8	100
		VAL	SP 2.2	100
		VER	GP 2.8	100
		VER	SP 3.2	100
	Control traceability	CM	SP 3.1	100
		REQM	SP 1.4	100
7.5.4	**Customer property**			
	Exercise care	PP	SP 2.3	30
	Identify property	PP	SP 2.3	30
	Report damage	PMC	SP 1.4	30
		PP	SP 2.3	
7.5.5	**Preservation of product**			
	Maintain conformity during delivery	PI	SP 3.4	100
	Preserve identification	PI	SP 3.4	100
	Preserve product component parts	PI	SP 3.4	100

(continued)

Table D.4 ISO 9001:2000 Section 7, Product Realization (continued)

Section	*ISO 9001:2000 Requirement Keywords*	*CMMI PAs*	*CMMI Practices*	*Confidence*	*Comment*
7.6	**Control of monitoring and measuring devices**				
	Determine monitoring and devices needed	MA	SP 1.1	100	
		MA	SP 1.2	100	
		MA	SP 1.3	100	
		MA	SP 1.4	100	
		VAL	SP 1.3	100	
		VER	SP 1.3	100	
	Establish monitoring processes	MA	GP 2.2	60	
		MA	GP 2.9	60	
		MA	SP 1.3	60	
		MA	SP 1.4	60	
		VAL	GP 2.9	60	

	VAL	SP 1.3	60
	VER	GP 2.9	60
	VER	SP 1.3	60
Calibrate at specified intervals			0
Adjust as needed			0
Calibration status			0
Safeguard from adjustment			0
Protect from damage			0
Assess prior measurement results			0
Take action on equipment			0
Keep calibration records			0
Confirm applicability of software	VAL	SP 1.2	30
	VAL	SP 1.3	30
	VER	SP 1.2	30
	VER	SP 1.3	30
Confirm software before use			0

Table D.5 ISO 9001:2000 Section 8, Measurement, Analysis and Improvement

Section	*ISO 9001:2000 Requirement Keywords*	*CMMI PAs*	*CMMI Practices*	*Confidence*	*Comment*
8	**Measurement, analysis and improvement**				
8.1	**General**				
	Product conformity	PPQA	SP 1.2	100	
		VER	GP 2.8	100	
		VER	SP 1.1	100	
		VER	SP 1.2	100	
		VER	SP 1.3	100	
		VER	SP 2.2	100	
		VER	SP 3.1	100	
	QMS conformity	OPF	SP 1.2	100	
	QMS improvement	OID	All SPs	100	
		OPF	SP 1.3	100	
		OPF	SP 2.1	100	
		OPF	SP 2.2	100	
	Determine methods and techniques	All PAs	GP 2.8	60	
		MA	SP 1.2	100	
		MA	SP 1.3	100	
		MA	SP 1.4	100	
		QPM	SP 2.1	100	
		QPM	SP 2.2	100	
8.2	**Monitoring and measurement**				
8.2.1	**Customer satisfaction**				
	Monitor customer perceptions	MA	SP 1.1	30	

		MA	SP 1.2	30	
		MA	SP 2.2	30	
		PMC	SP 1.5	30	
		VAL	GP 2.7	60	
		VAL	SP 2.1	60	
	Define methods for measuring satisfaction	MA	SP 1.2	60	
		MA	SP 1.3	60	
		MA	SP 1.4	60	
8.2.2	**Internal audit**				
	Determine conformance	OPF	SP 1.2	100	
		PPQA	All SPs	100	
	Determine effectiveness	OPF	SP 1.1	100	
		OPF	SP 1.2	100	
		OPF	SP 1.3	100	
	Audits consider process importance	OPF	SP 1.1	100	
		OPF	SP 1.2	100	
		PPQA	GP 2.2	100	
	Define audit criteria	OPF	SP 1.1	100	
		OPF	SP 1.2	100	
		PPQA	GP 2.2	100	
	Select objective auditors	PPQA	GP 2.4	100	Objectivity is addressed in PPQA introductory notes.
	Don't audit own output			0	Objectivity is addressed in PPQA introductory notes.
	Documented procedure defines audits	OPF	GP 2.2	100	

(continued)

Table D.5 ISO 9001:2000 Section 8, Measurement, Analysis and Improvement (continued)

Section	*ISO 9001:2000 Requirement Keywords*	*CMMI PAs*	*CMMI Practices*	*Confidence*	*Comment*
		OPF	GP 2.4	100	
		PPQA	GP 2.2	100	
		PPQA	GP 2.4	100	
		PPQA	SP 2.2	100	
	Actions taken promptly	OPF	GP 2.1	100	
		OPF	GP 2.4	100	
		PPQA	GP 2.1	100	
		PPQA	GP 2.4	100	
		PPQA	SP 2.1	100	
	Verify actions taken	OPF	SP 2.1	30	
		OPF	SP 2.2	60	
		OPF	SP 3.3	100	
		PPQA	SP 2.1	100	
		PPQA	SP 2.2	100	
8.2.3	**Monitoring and measurement of process**				
	Use suitable methods	All PAs	GP 2.8	100	
		MA	SP 1.2	100	
		MA	SP 1.3	100	
		MA	SP 1.4	100	
		OPP	SP 1.2	100	

	Demonstrate process capability	MA	SP 2.2	100
		QPM	SP 2.1	100
		QPM	SP 2.2	100
		QPM	SP 2.3	100
		QPM	SP 2.4	100
	Take corrective actions	PMC	SP 2.1	100
		PMC	SP 2.2	100
		PMC	SP 2.3	100
		QPM	SP 2.3	100
8.2.4	**Monitoring and measurement of product**			
	Monitor product characteristics	All PAs	GP 2.8	60
		MA	SP 2.1	100
		MA	SP 2.2	100
		VAL	SP 2.1	100
		VAL	SP 2.2	100
		VER	SP 3.1	100
		VER	SP 3.2	100
	Measure at appropriate stages	VAL	All SPs	100
	All SPs	VER	All SPs	100
	Maintain conformity record	PI	SP 3.3	100
		SAM	SP 2.4	100
		VAL	GP 2.6	60
		VER	GP 2.6	60
	Maintain release records	PI	GP 2.6	60
	Don't release until product realization plans are implemented	CM	SP 3.2	30

(continued)

Table D.5 ISO 9001:2000 Section 8, Measurement, Analysis and Improvement (continued)

Section	*ISO 9001:2000 Requirement Keywords*	*CMMI PAs*	*CMMI Practices*	*Confidence*	*Comment*
8.3	**Control of nonconforming product**				
	Identify and control nonconforming product	CM	SP 1.3	60	
		PMC	SP 2.1	100	
		PMC	SP 2.2	100	
		PMC	SP 2.3	100	
		VAL	SP 2.2	60	
		VER	SP 3.2	60	
	Define control of nonconforming product	VAL	GP 2.2	100	
		VAL	GP 2.4	100	
		VER	GP 2.2	100	
		VER	GP 2.4	100	
	Take action	CM	SP 2.1	100	
		CM	SP 2.2	100	
		PPQA	SP 2.1	100	
		PMC	SP 2.1	100	
		PMC	SP 2.2	100	
		PMC	SP 2.3	100	
		VER	SP 3.2	100	
	Authorize use	PMC	SP 2.1	100	
		PMC	SP 2.2	100	
		PMC	SP 2.3	100	

		VER	SP 3.2	100	
	Preclude use	PMC	SP 2.1	100	
		PMC	SP 2.2	100	
		PMC	SP 2.3	100	
		VER	SP 3.2	100	
	Keep records of nonconformities	CM	SP 3.1	100	
		VER	GP 2.6	60	
		VER	SP 3.2	100	
	Reverify corrected nonconformance			0	Implied by the nonsequential and recursive nature of CMMI
	Take action after delivery			0	
8.4	**Analysis of data**				
	Collect data on QMS effectiveness	All PAs	GP 3.2	100	
		IPM	SP 1.6	100	
		MA	All SPs	100	
		OID	SP 1.1	100	
		OID	SP 1.2	100	
		OID	SP 1.3	100	
		OID	SP 1.4	100	
		OPF	SP 1.2	100	
		OPF	SP 1.3	100	
		OPP	SP 1.1	100	
		OPP	SP 1.2	100	
		OPP	SP 1.3	100	
	Include monitoring data	All PAs	GP 2.8	100	

(continued)

Table D.5 ISO 9001:2000 Section 8, Measurement, Analysis and Improvement (continued)

Section	*ISO 9001:2000 Requirement Keywords*	*CMMI PAs*	*CMMI Practices*	*Confidence*	*Comment*
	Analyze customer satisfaction	MA	SP 1.1	30	
		MA	SP 1.2	30	
		MA	SP 2.2	30	
		PMC	SP 1.5	30	
		VAL	GP 2.7	30	
		VAL	SP 2.1	30	
	Analyze conformance	VER	SP 3.2	100	
	Analyze trends	CAR	SP 1.1	60	
		CAR	SP 1.2	60	
		OPP	SP 1.4	100	
		QPM	SP 1.4	100	
	Analyze suppliers	SAM	GP 2.8	60	
		SAM	SP 2.2	100	
		SAM	SP 2.3	100	
8.5	**Improvement**				
8.5.1	**Continual improvement**				
	Improve QMS effectiveness	All PAs	GP 2.1	100	
		All PAs	GP 2.10	100	
		CAR	All SPs	100	
		MA	SP 1.1	100	
		MA	SP 2.2	100	
		OID	All SPs	100	
		OPF	All SPs	100	

8.5.2	**Corrective action**			
	Eliminate causes of nonconformities	CAR	All SPs	100
		OPF	SP 2.1	100
		OPF	SP 2.2	100
		OPF	SP 3.1	100
	Take appropriate actions			0
	Review nonconformities	CAR	GP 2.2	60
		CAR	SP 1.1	100
		PMC	GP 2.2	60
		PMC	SP 2.1	100
		PPQA	SP 2.1	100
		VER	GP 2.2	60
		VER	SP 3.2	100
	Determine causes	CAR	SP 1.2	100
	Evaluate need for action	CAR	SP 1.2	100
		PMC	SP 2.1	100
	Determine action needed	CAR	GP 2.2	100
		CAR	SP 1.1	100
		CAR	SP 1.2	100
		CAR	SP 2.1	100
		PMC	SP 2.2	100
		PPQA	SP 2.1	100
	Record results	CAR	SP 2.3	100
		PMC	GP 2.6	60
		PPQA	SP 2.2	100

(continued)

Table D.5 ISO 9001:2000 Section 8, Measurement, Analysis and Improvement (continued)

Section	*ISO 9001:2000 Requirement Keywords*	*CMMI PAs*	*CMMI Practices*	*Confidence*	*Comment*
	Review action	CAR	SP 2.2	100	
		PMC	SP 2.3	100	
		PPQA	SP 2.1	100	
8.5.3	**Preventive action**				
	Determine action to prevent nonconformity	CAR	SP 1.1	100	
		CAR	SP 1.2	100	
		OPF	SP 3.4	100	
	Take appropriate actions			0	
	Determine potential nonconformities	CAR	GP 2.2	60	
		CAR	SP 1.1	100	
		PMC	GP 2.2	60	

	PMC	SP 2.1	100
	PPQA	SP 2.1	100
	VER	GP 2.2	60
	VER	SP 3.2	100
Evaluate need for action	CAR	SP 1.2	100
	PMC	SP 2.1	100
Determine action needed	CAR	GP 2.2	100
	CAR	SP 1.1	100
	CAR	SP 1.2	100
	CAR	SP 2.1	100
	PMC	SP 2.2	100
Record results	CAR	SP 2.3	100
	PMC	GP 2.6	60
Review action	CAR	SP 2.2	100
	PMC	SP 2.3	100

Appendix E: ISO 15288:2008 to CMMI v1.2 Map

Table E.1 ISO 15288:2008 Section 6.1, Agreement Processes

Section	*ISO 15288 Keywords*	*CMMI PAs*	*CMMI Practices*	*Confidence*	*Comment*
6	**System Life Cycle Processes**				
6.1	**Agreement Processes**				
6.1.1	**Acquisition Process**				
6.1.1.3	**Activities and tasks**				
6.1.1.3a	**Prepare for the acquisition**				
6.1.1.3a 1	Establish acquisition strategy	SAM	SP 1.1	30	
6.1.1.3a 2	Prepare request	SAM	SP 1.2	30	
6.1.1.3b	**Advertise the acquisition and select the supplier**				
6.1.1.3b 1	Communicate request	SAM	SP 1.2	60	
6.1.1.3b 2	Select supplier	SAM	SP 1.2	100	
6.1.1.3c	**Initiate an agreement**				
6.1.1.3c 1	Negotiate agreement	SAM	SP 1.3	60	
6.1.1.3c 2	Commence agreement	SAM	SP 2.1	60	
6.1.1.3d	**Monitor the agreement**				
6.1.1.3d 1	Assess execution	SAM	SP 2.1	100	
		SAM	SP 2.2	100	
		SAM	SP 2.3	100	
6.1.1.3d 2	Provide data	SAM	SP 2.1	100	
6.1.1.3e	**Accept the product or service**				
6.1.1.3e 1	Confirm compliance	SAM	SP 2.4	100	

6.1.1.3e 2	Make payment			0	
6.1.2	**Supply Process**				
6.1.2.3	**Activities and tasks**				
6.1.2.3a	**Identify opportunities**			0	Marketing activities not addressed in CMMI
6.1.2.3b	**Respond to a tender**				
6.1.2.3b 1	Evaluate request			0	
6.1.2.3b 2	Prepare response			0	Proposal activities not addressed in CMMI
6.1.2.3c	**Initiate an agreement**				
6.1.2.3c 1	Negotiate agreement	PP	SP 3.3	60	Contractual aspects of agreement not explicitly addressed in CMMI
		REQM	SP 1.2	60	
6.1.2.3c 2	Commence agreement			0	
6.1.2.3d	**Execute the agreement**				
6.1.2.3d 1	Execute agreement	IPM	SP 1.5	100	
		PMC	All SPs	100	
6.1.2.3d 2	Assess execution	PMC	All SPs	100	
6.1.2.3e	**Deliver and support the product or service**				
6.1.2.3e 1	Deliver the product	PI	SP 3.4	100	
6.1.2.3e 2	Provide assistance			0	
6.1.2.3f	**Close the agreement**				
6.1.2.3f 1	Accept payment			0	
6.1.2.3f 2	Transfer responsibility	PI	SP 3.4	30	

Table E.2 ISO 15288:2008 Section 6.2, Organizational Project-Enabling Processes

Section	*ISO 15288 Keywords*	*CMMI PAs*	*CMMI Practices*	*Confidence*	*Comment*
6.2	**Organizational Project-Enabling Processes**				
6.2.1	**Life Cycle Model Management Process**				
6.2.1.3	**Activities and tasks**				
6.2.1.3a	**Establish the process**				
6.2.1.3a 1	Establish policies and procedures	OPD	GP 2.1	100	
		OPD	GP 2.2	100	
		OPF	GP 2.1	100	
		OPF	GP 2.2	100	
6.2.1.3a 2	Establish processes	OPD	SP 1.1	100	
		OPD	SP 1.2	100	
		OPD	SP 1.3	100	
6.2.1.3a 3	Define roles	OPD	GP 2.4	100	
		OPF	GP 2.4	100	
6.2.1.3a 4	Life cycle progression	OPD	SP 1.1	100	
6.2.1.3a 5	Establish life cycle models	OPD	SP 1.2	100	
6.2.1.3b	**Assess the process**				
6.2.1.3b 1	Monitor execution	All	GP 2.8	60	No specific CMMI requirements for business criteria review or effectiveness and efficiency feedback
		OPP	SP 1.4	100	
6.2.1.3b 2	Review life cycle model	OPF	SP 1.2	30	No specific requirement for LCM review
		OPF	SP 3.3	30	
6.2.1.3b 3	Identify improvements	OID	SP 1.1	60	
		OID	SP 1.2	60	

		OPF	SP 1.3	100
6.2.1.3c	**Improve the process**			
6.2.1.3c 1	Prioritize	OPF	SP 2.1	100
6.2.1.3c 2	Implement	OID	SP 1.3	100
		OID	SP 1.4	100
		OPF	SP 3.4	100
6.2.2	**Infrastructure Management Process**			
6.2.2.3	**Activities and tasks**			
6.2.2.3a	**Establish the infrastructure**			
6.2.2.3a 1	Define requirements	All PAs	GP 2.3	100
		OPD	SP 1.6	100
6.2.2.3a 2	Identify and provide resources	IPM	SP 1.3	100
		OPD	SP 1.6	100
		PP	SP 2.4	100
		All	GP 2.3	100
6.2.2.3b	**Maintain the infrastructure**			
6.2.2.3b 1	Determine project satisfaction			0
6.2.2.3b 2	Improve with project changes			0
6.2.3	**Project Portfolio Management Process**			
6.2.3.3	**Activities and tasks**			
6.2.3.3a	**Initiate projects**			
6.2.3.3a 1	Prioritize opportunities			0
6.2.3.3a 2	Define projects	IPM	GP 2.4	100
		PMC	GP 2.4	100
		PP	GP 2.4	100
		PP	SP 2.7	100

(continued)

Table E.2 ISO 15288:2008 Section 6.2, Organizational Project-Enabling Processes (continued)

Section	*ISO 15288 Keywords*	*CMMI PAs*	*CMMI Practices*	*Confidence*	*Comment*
6.2.3.3a 3	Identify goals			0	
6.2.3.3a 4	Allocate resources	All PAs	GP 2.3	100	
6.2.3.3a 5	Identify dependencies	OPD	SP 1.6	30	
		PP	SP 2.6	60	
6.2.3.3a 6	Specify reporting	PP	SP 2.6	30	
		PP	SP 2.7	60	
		PP	GP 2.4	100	
6.2.3.3a 7	Authorize project			0	
6.2.3.3b	**Evaluate the portfolio of projects**				
6.2.3.3b 1	Evaluate projects	IPM	GP 2.10	100	
		PMC	GP 2.10	100	
		PPQA	GP 2.10	100	
		PP	GP 2.10	100	
6.2.3.3b 2	Continue projects	IPM	GP 2.10	60	
		PMC	GP 2.10	60	
		PPQA	GP 2.10	60	
		PP	GP 2.10	60	
6.2.3.3c	**Close projects**				
6.2.3.3c 1	Cancel projects	PP	GP 2.10	60	
		PMC	GP 2.10	60	
		IPM	GP 2.10	60	
		RSKM	GP 2.10	60	
6.2.3.3c 2	Close projects			0	

6.2.4	**Human Resource Management Process**				
6.2.4.3	**Activities and tasks**				
6.2.4.3a	**Identify skills**				
6.2.4.3a 1	Identify needs	OT	SP 1.1	100	
6.2.4.3a 2	Identify skills	OT	SP 1.1	100	
6.2.4.3b	**Develop skills**				
6.2.4.3b 1	Plan skill development	OT	SP 1.3	100	
6.2.4.3b 2	Obtain training resources	OT	SP 1.4	100	
6.2.4.3b 3	Provide development	OT	SP 2.1	100	
6.2.4.3b 4	Maintain records	OT	SP 2.2	100	
6.2.4.3c	**Acquire and provide skills**				
6.2.4.3c 1	Obtain personnel	PMC	SP 1.1	60	
6.2.4.3c 2	Manage personnel skills	OT	SP 1.1	30	
		OT	SP 1.3	30	
		OT	SP 2.1	100	
6.2.4.3c 3	Make assignments	PP	SP 2.5	100	
6.2.4.3c 4	Motivate personnel			0	
6.2.4.3c 5	Control project interfaces	IPM	SP 2.2	30	
		IPM	SP 2.3	30	
		IPM	SP 3.5	30	
6.2.4.3d	**Perform knowledge management**				
6.2.4.3d 1	Establish infrastructure	OT	SP 1.4	30	
6.2.4.3d 2	Knowledge management strategy	OT	SP 1.3	30	Strategy not addressed in CMMI
6.2.4.3d 3	Maintain information	OT	SP 1.3	30	
6.2.5	**Quality Management Process**				
6.2.5.3	**Activities and tasks**				

(continued)

Table E.2 ISO 15288:2008 Section 6.2, Organizational Project-Enabling Processes (continued)

Section	*ISO 15288 Keywords*	*CMMI PAs*	*CMMI Practices*	*Confidence*	*Comment*
6.2.5.3a	**Plan quality management**				
6.2.5.3a 1	Establish policies and procedures	OPF	GP 2.1	100	QMS is equivalent to OSSP
		OPF	GP 2.2	100	
		OPD	GP 2.1	100	
		OPD	GP 2.2	100	
		OPD	SP 1.1	100	
6.2.5.3a 2	Establish objectives	MA	SP 1.1	60	Must specify "customer satisfaction"
		OPF	SP 1.1	60	
		OPP	SP 1.3	60	
		QPM	SP 1.1	60	
6.2.5.3a 3	Define responsibilities	MA	GP 2.4	100	
		OPF	GP 2.4	100	
		OPP	GP 2.4	100	
		PPQA	GP 2.4	100	

		QPM	GP 2.4	100	
		OPD	GP 2.4	100	
6.2.5.3b	**Assess quality management**				
6.2.5.3b 1	Assess satisfaction	MA	SP 2.2	30	Weak and not explicit in CMMI
		MA	SP 2.4	30	
6.2.5.3b 2	Review plans	All PAs	GP 2.9	60	
		All PAs	GP 2.10	60	
		PP	SP 3.1	100	
		IPM	SP 1.4	100	
6.2.5.3b 3	Monitor status	All PAs	GP 2.8	30	
		OID	SP 2.3	100	
		OPF	SP 2.2	30	
		QPM	SP 2.3	100	
6.2.5.3c	**Perform quality management corrective action**				
6.2.5.3c 1	Plan corrective action	PPQA	SP 2.1	60	Planning is not explicit
6.2.5.3c 2	Implement actions	PPQA	SP 2.2	100	Organizational communication not addressed

Table E.3 ISO 15288:2008 Section 6.3, Project Processes

Section	*ISO 15288 Keywords*	*CMMI PAs*	*CMMI Practices*	*Confidence*	*Comment*
6.3	**Project Processes**				
6.3.1	**Project Planning Process**				
6.3.1.3	**Activities and tasks**				
6.3.1.3a	**Define the project**				
6.3.1.3a 1	Identify objectives	IPM	SP 1.1	30	PP and IPM focus on constraints rather than objectives
		PP	SP 2.1	60	
		PP	SP 3.2	60	
		QPM	SP 1.1	100	
6.3.1.3a 2	Define scope	PP	SP 1.1	100	
6.3.1.3a 3	Define life cycle model	PP	SP 1.3	100	
		IPM	SP 1.1	100	
6.3.1.3a 4	Establish Work Breakdown Structure	PP	SP 1.1	100	
6.3.1.3b	**Plan the project resources**				
6.3.1.3b 1	Define schedule	IPM	SP 1.4	100	
		IPM	SP 2.2	100	
		PP	SP 1.2	100	
		PP	SP 1.4	100	
		PP	SP 2.1	100	
6.3.1.3b 2	Define criteria	IPM	SP 1.1	100	
		IPM	SP 1.4	100	
		PP	SP 1.3	100	
		PP	SP 2.1	100	
6.3.1.3b 3	Define budget	IPM	SP 1.2	30	

		PP	SP 1.4	100	
		PP	SP 2.1	100	
6.3.1.3b 4	Establish responsibilities	All PAs	GP 2.4	100	
		IPM	SP 3.4	100	
		PP	SP 2.4	100	
6.3.1.3b 5	Define project infrastructure	IPM	SP 1.3	100	
		PP	GP 2.3	100	
		PP	SP 2.4	100	
6.3.1.3b 6	Plan acquisitions	IPM	SP 1.3	60	
		PP	SP 2.4	100	
		SAM	GP 2.2	100	
6.3.1.3c	**Plan the project technical and quality management**				
6.3.1.3c 1	Generate management plan	IPM	SP 1.4	100	
		PP	SP 2.7	100	
		PP	SP 3.3	60	
6.3.1.3c 2	Generate quality plan	IPM	SP 1.4	100	
		PP	SP 2.7	100	
6.3.1.3d	**Activate the project**				
6.3.1.3d 1	Authorize project			0	
6.3.1.3d 2	Commit resources	PP	SP 3.3	100	
6.3.1.3d 3	Initiate project	IPM	SP 1.5	60	"Initiation" not specifically addressed
6.3.2	**Project Assessment and Control Process**				
6.3.2.3	**Activities and tasks**				
6.3.2.3a	**Assess the project**				

(continued)

Table E.3 ISO 15288:2008 Section 6.3, Project Processes (continued)

Section	*ISO 15288 Keywords*	*CMMI PAs*	*CMMI Practices*	*Confidence*	*Comment*
6.3.2.3a 1	Assess project status	IPM	SP 1.5	100	
		PMC	SP 1.1	100	
		PMC	SP 1.2	100	
		PMC	SP 1.3	100	
		PMC	SP 1.4	100	
		PMC	SP 1.5	100	
		PMC	SP 1.6	100	
		PMC	SP 1.7	100	
6.3.2.3a 2	Perform Quality Assurance	PPQA	SP 1.1	100	
		PPQA	SP 1.2	100	
		PPQA	SP 2.1	100	
		PPQA	SP 2.2	100	
6.3.2.3a 3	Assess team effectiveness	IPM	GP 2.8	100	Implies IPPD addition
6.3.2.3a 4	Assess infrastructure adequacy	PMC	SP 1.1	60	
		PMC	SP 1.2	100	
6.3.2.3a 5	Assess progress	IPM	SP 1.5	100	
		PMC	SP 1.1	100	
		PMC	SP 1.6	100	
		PMC	SP 1.7	100	
6.3.2.3a 6	Conduct reviews and audits	PMC	SP 1.6	100	
		PMC	SP 1.7	100	
		IPM	SP 1.5	100	
6.3.2.3a 7	Monitor critical processes	QPM	SP 2.3	60	

		IPM	SP 1.5	30	
6.3.2.3a 8	Analyze measures	All PAs	GP 2.8	100	
		IPM	SP 1.5	100	
		PMC	All SPs	100	
6.3.2.3a 9	Report status	PMC	SP 1.6	100	
		PMC	SP 1.6	100	
		PMC	SP 2.3	100	
6.3.2.3b	**Control the project**				
6.3.2.3b 1	Manage requirements	REQM	SP 1.2	100	
		REQM	SP 1.3	100	
6.3.2.3b 2	Initiate corrective action	PMC	SP 2.2	100	
6.3.2.3b 3	Initiate preventive action	CAR	SP 1.1	60	
		CAR	SP 1.2	60	
		CAR	SP 2.1	60	
6.3.2.3b 4	Initiate problem resolution	PMC	SP 2.2	100	
		PPQA	SP 2.1	100	
6.3.2.3b 5	Evolve scope of work	PP	SP 1.1	100	
		REQM	SP 1.5	100	
6.3.2.3b 6	Initiate changes	REQM	SP 1.5	100	
		SAM	SP 1.3	100	
6.3.2.3b 7	Correct defective provision	PMC	SP 2.2	100	
		SAM	SP 2.3	100	
		SAM	SP 2.4	100	
6.3.2.3b 8	Authorize project to proceed	PMC	SP 1.7	30	"Authorize" not explicit in CMMI
6.3.2.3c	**Close the project**				
6.3.2.3c 1	Determine project completion			0	Nothing about "closure" in CMMI

(continues)

Table E.3 ISO 15288:2008 Section 6.3, Project Processes (continued)

Section	*ISO 15288 Keywords*	*CMMI PAs*	*CMMI Practices*	*Confidence*	*Comment*
6.3.2.3c 2	Archive results			0	Nothing about "archiving" in CMMI
6.3.3	**Decision Management Process**				
6.3.3.3	**Activities and tasks**				
6.3.3.3a	**Plan and define decisions**				
6.3.3.3a 1	Define strategy	DAR	GP 2.2	100	
		DAR	SP 1.1	100	
6.3.3.3a 2	Identify need for decision	DAR	SP 1.1	100	
		DAR	SP 1.2	100	
		DAR	SP 1.3	100	
6.3.3.3a 3	Involve relevant parties	DAR	GP 2.7	100	
6.3.3.3b	**Analyze the decision information**				
6.3.3.3b 1	Select strategy	DAR	SP 1.4	100	
6.3.3.3b 2	Identify success criteria	DAR	SP 1.4	100	
6.3.3.3b 3	Evaluate alternatives	DAR	SP 1.5	100	
6.3.3.3c	**Track the decision**				
6.3.3.3c 1	Record decisions	DAR	SP 1.6	100	
6.3.3.3c 2	Maintain records	DAR	GP 2.6	100	
		DAR	SP 1.6	100	
6.3.4	**Risk Management Process**				
6.3.4.3	**Activities and tasks**				
6.3.4.3a	**Plan risk management**				
6.3.4.3a 1	Define policies	RSKM	GP 2.1	100	
6.3.4.3a 2	Document risk process	RSKM	GP 2.2	100	

		RSKM	GP 3.1	100	
6.3.4.3a 3	Identify responsible parties	RSKM	GP 2.4	100	
		RSKM	GP 2.7	100	
6.3.4.3a 4	Provide resources	RSKM	GP 2.3	100	
6.3.4.3a 5	Define risk improvement process	RSKM	GP 2.8	30	
		RSKM	GP 3.2	60	
6.3.4.3b	**Manage the risk profile**				
6.3.4.3b 1	Document risk context	RSKM	SP 1.1	100	
6.3.4.3b 2	Define thresholds	RSKM	SP 1.2	100	
6.3.4.3b 3	Establish risk profile	RSKM	SP 2.2	100	
6.3.4.3b 4	Communicate risk profile	RSKM	GP 2.7	100	
6.3.4.3c	**Analyze risks**				
6.3.4.3c 1	Identify risks	PP	SP 2.2	100	
		RSKM	SP 2.1	100	
6.3.4.3c 2	Estimate probability	RSKM	SP 2.2	100	
6.3.4.3c 3	Evaluate risks	RSKM	SP 2.2	100	
6.3.4.3c 4	Define risk treatment strategy	RSKM	SP 3.1	100	
6.3.4.3d	**Treat risks**				
6.3.4.3d 1	Recommend risk treatment	RSKM	SP 3.1	100	
6.3.4.3d 2	Implement risk treatment	RSKM	SP 3.2	100	
6.3.4.3d 3	Monitor risks	RSKM	SP 3.2	100	
6.3.4.3d 4	Manage in accordance with section 6.3.2.3				See section 6.3.2.3
6.3.4.3e	**Monitor risks**				
6.3.4.3e 1	Monitor risks and context	PMC	SP 1.3	100	
		RSKM	SP 3.2	100	
6.3.4.3e 2	Monitor measures	RSKM	GP 2.8	100	

(continued)

Table E.3 ISO 15288:2008 Section 6.3, Project Processes (continued)

Section	*ISO 15288 Keywords*	*CMMI PAs*	*CMMI Practices*	*Confidence*	*Comment*
6.3.4.3e 3	Monitor new risks	PP	SP 2.1	100	
		RSKM	SP 2.1	100	
6.3.4.3f	**Evaluate the Risk Management Process**				
6.3.4.3f 1	Collect improvement information	RSKM	GP 3.2	100	
6.3.4.3f 2	Review risk management process	RSKM	GP 2.8	30	Effectiveness and efficiency not explicit
6.3.4.3f 3	Review risk information	OPF	SP 3.4	100	This should be specifically addressed in reviews
		RSKM	GP 3.2	60	
		RSKM	GP 2.10	60	
6.3.5	**Configuration Management Process**				
6.3.5.3	**Activities and tasks**				
6.3.5.3a	**Plan configuration management**				
6.3.5.3a 1	Define CM strategy	CM	GP 2.2	100	
		CM	SP 1.2	100	
6.3.5.3a 2	Identify items	CM	SP 1.1	100	
6.3.5.3b	**Perform configuration management**				
6.3.5.3b 1	Maintain configuration information	CM	SP 1.3	100	
		CM	SP 2.1	100	
		CM	SP 3.1	100	
6.3.5.3b 2	Maintain baselines	CM	SP 2.1	100	
		CM	SP 2.2	100	
		CM	SP 3.1	100	
		CM	SP 3.2	100	

6.3.6	**Information Management Process**				
6.3.6.3	**Activities and tasks**				
6.3.6.3a	**Plan information management**				
6.3.6.3a 1	Define items to be managed	PP	SP 2.3	100	
6.3.6.3a 2	Designate responsibilities	PP	SP 2.3	100	
6.3.6.3a 3	Define rights	PP	SP 2.3	100	
6.3.6.3a 4	Define representation	PP	SP 2.3	100	
6.3.6.3a 5	Define maintenance	PP	SP 2.3	100	
6.3.6.3b	**Perform information management**				
6.3.6.3b 1	Obtain information items	PMC	SP 1.4	30	Data collection activities are not explicitly identified in CMMI
6.3.6.3b 2	Maintain information items	PMC	SP 1.4	100	
6.3.6.3b 3	Retrieve information	PMC	SP 1.4	30	
6.3.6.3b 4	Provide documentation	PMC	SP 1.4	30	
6.3.6.3b 5	Archive information	PMC	SP 1.4	30	
6.3.6.3b 6	Dispose of information	PMC	SP 1.4	30	
6.3.7	**Measurement Process**				
6.3.7.3	**Activities and tasks**				
6.3.7.3a	**Plan measurement**				
6.3.7.3a 1	Describe relevant characteristics	MA	SP 1.1	100	
6.3.7.3a 2	Identify information needs	MA	SP 1.1	100	
6.3.7.3a 3	Select measures	MA	SP 1.2	100	
6.3.7.3a 4	Define procedures	MA	SP 1.3	100	
6.3.7.3a 5	Define evaluation criteria	MA	SP 1.4	100	
6.3.7.3a 6	Provide resources	MA	GP 2.3	100	
6.3.7.3a 7	Deploy technologies	MA	GP 2.3	100	

(continued)

Table E.3 ISO 15288:2008 Section 6.3, Project Processes (continued)

Section	*ISO 15288 Keywords*	*CMMI PAs*	*CMMI Practices*	*Confidence*	*Comment*
6.3.7.3b	**Perform measurement**				
6.3.7.3b 1	Integrate procedures	MA	GP 2.2	60	Integration is not explicit in CMMI
6.3.7.3b 2	Collect data	MA	SP 2.1	100	
		MA	SP 2.3	100	
6.3.7.3b 3	Analyze data	MA	SP 2.2	100	
6.3.7.3b 4	Document results	MA	SP 2.4	100	
6.3.7.3c	**Evaluate measurement**				
6.3.7.3c 1	Evaluate products and process	MA	GP 2.9	100	
6.3.7.3c 2	Identify improvements	MA	GP 3.2	100	

Table E.4 ISO 15288:2008 Section 6.4, Technical Processes

Section	*ISO 15288 Keywords*	*CMMI PAs*	*CMMI Practices*	*Confidence*	*Comment*
6.4	**Technical Processes**				
6.4.1	**Stakeholder Requirements Definition Process**				
6.4.1.3	**Activities and tasks**				
6.4.1.3a	**Elicit stakeholder requirements**				
6.4.1.3a 1	Identify stakeholders	All PAs	GP 2.7	100	
		PP	SP 2.6	100	
6.4.1.3a 2	Elicit requirements	RD	SP 1.1	100	
6.4.1.3b	**Define stakeholder requirements**				
6.4.1.3b 1	Define constraints	RD	SP 1.1	100	
		RD	SP 1.2	100	
6.4.1.3b 2	Define operational scenarios	RD	SP 3.1	100	
6.4.1.3b 3	Identify interactions	RD	SP 3.1	100	
6.4.1.3b 4	Specify environmental requirements	RD	SP 1.1	60	
		RD	SP 1.2	60	
		RD	SP 2.1	60	
		RD	SP 2.3	100	
		RD	SP 3.2	60	
6.4.1.3c	**Analyze and maintain stakeholder requirements**				
6.4.1.3c 1	Analyze requirements	RD	SP 3.3	100	
6.4.1.3c 2	Resolve problems	RD	SP 3.3	100	
		RD	SP 3.4	30	

(continued)

Table E.4 ISO 15288:2008 Section 6.4, Technical Processes (continued)

Section	*ISO 15288 Keywords*	*CMMI PAs*	*CMMI Practices*	*Confidence*	*Comment*
6.4.1.3c 3	Feedback requirements	RD	SP 3.5	100	
6.4.1.3c 4	Confirm requirements understanding	RD	SP 3.5	100	
6.4.1.3c 5	Record requirements	RD	SP 1.2	100	
		RD	SP 2.2	100	
		REQM	GP 2.6	100	
		REQM	SP 1.3	100	
6.4.1.3c 6	Maintain traceability	REQM	SP 1.4	100	
6.4.2	**Requirements Analysis Process**				
6.4.2.3	**Activities and tasks**				
6.4.2.3a	**Define system requirements**				
6.4.2.3a 1	Define system boundary	RD	SP 2.3	100	
		RD	SP 3.1	100	
6.4.2.3a 2	Define system functions	RD	SP 3.2	100	
6.4.2.3a 3	Define constraints	RD	SP 1.1	100	
		RD	SP 2.2	100	
		TS	SP 1.2	30	
6.4.2.3a 4	Define measures	RD	SP 3.3	60	
		MA	SP 1.2	60	
6.4.2.3a 5	Specify system requirements	RD	SP 1.2	60	
		RD	SP 2.1	60	
		RD	SP 3.3	60	
		RD	SP 3.4	60	
		RD	SP 3.5	60	

6.4.2.3b	**Analyze and maintain system requirements**				
6.4.2.3b 1	Analyze requirements	RD	SP 3.3	100	
		RD	SP 3.5	100	
6.4.2.3b 2	Feedback requirements	RD	SP 3.5	100	
		RD	GP 2.7	100	
6.4.2.3b 3	Demonstrate traceability	REQM	SP 1.4	100	
6.4.2.3b 4	Maintain requirements	RD	SP 1.2	100	
		RD	SP 2.1	100	
6.4.3	**Architectural Design Process**				
6.4.3.3	**Activities and tasks**				
6.4.3.3a	**Define the architecture**				
6.4.3.3a 1	Define logical architecture	RD	SP 2.2	100	
		RD	SP 3.2	100	
		TS	SP 1.1	100	
		TS	SP 1.2	100	
6.4.3.3a 2	Partition functions	RD	SP 2.2	100	
		RD	SP 3.2	100	
6.4.3.3a 3	Document interfaces	RD	SP 2.3	100	
		TS	SP 2.3	100	
6.4.3.3b	**Analyze and evaluate the architecture**				
6.4.3.3b 1	Analyze architecture	TS	SP 1.1	100	
		TS	SP 2.1	100	
6.4.3.3b 2	Allocate requirements to operators	RD	SP 2.2	30	ISO 15288 is more detailed than CMMI
		RD	SP 2.3	30	
6.4.3.3b 3	Determine available off-the-shelf elements	TS	SP 1.1	100	
		TS	SP 1.2	100	

(continued)

Table E.4 ISO 15288:2008 Section 6.4, Technical Processes (continued)

Section	*ISO 15288 Keywords*	*CMMI PAs*	*CMMI Practices*	*Confidence*	*Comment*
		TS	SP 2.4	100	
6.4.3.3b 4	Evaluate alternative designs	TS	SP 1.1	100	
		TS	SP 1.2	100	
6.4.3.3c	**Document and maintain the architecture**				
6.4.3.3c 1	Specify physical design	TS	SP 1.2	100	
		TS	SP 2.1	100	
6.4.3.3c 2	Record design	TS	SP 2.1	100	
		TS	SP 2.2	100	
6.4.3.3c 3	Maintain traceability	REQM	SP 1.4	100	
6.4.4	**Implementation Process**				
6.4.4.3	**Activities and tasks**				
6.4.4.3a	**Plan the implementation**				
6.4.4.3a 1	Generate strategy	TS	GP 2.2	100	
6.4.4.3a 2	Identify constraints	TS	SP 3.1	30	
6.4.4.3b	**Perform implementation**				
6.4.4.3b 1	Realize system elements	TS	SP 3.1	100	
6.4.4.3b 2	Record evidence	PPQA	SP 1.2	100	
		SAM	SP 2.3	60	
		TS	SP 3.1	100	
		VER	SP 3.2	100	
6.4.4.3b 3	Package system elements	PI	SP 3.4	100	
6.4.5	**Integration Process**				
6.4.5.3	**Activities and tasks**				

6.4.5.3a	**Plan integration**				
6.4.5.3a 1	Define assembly sequence	PI	SP 1.1	100	
6.4.5.3a 2	Identify constraints	PI	SP 1.1	60	
6.4.5.3b	Perform integration				
6.4.5.3b 1	Obtain enabling systems	PI	SP 1.2	100	
6.4.5.3b 2	Obtain system elements	PI	SP 3.1	100	
6.4.5.3b 3	Assure verification of system elements	PI	SP 3.1	100	
6.4.5.3b 4	Integrate system elements	PI	SP 2.2	100	
		PI	SP 3.2	100	
6.4.5.3b 5	Analyze integration information	PI	SP 3.3	60	CMMI does not address nonconformances and corrective actions
6.4.6	**Verification Process**				
6.4.6.3	**Activities and tasks**				
6.4.6.3a	**Plan verification**				
6.4.6.3a 1	Define strategy	VER	SP 1.1	100	
		VER	SP 1.2	100	
		VER	SP 1.3	100	
6.4.6.3a 2	Define verification plan	VER	GP 2.2	100	
		VER	SP 1.3	100	
6.4.6.3a 3	Identify constraints	TS	SP 2.1	60	
6.4.6.3b	**Perform verification**				
6.4.6.3b 1	Verification systems available	VER	SP 1.2	100	
6.4.6.3b 2	Conduct verification	VER	SP 2.2	100	
		VER	SP 3.1	100	
6.4.6.3b 3	Make data available	VER	SP 3.2	100	

(continued)

Table E.4 ISO 15288:2008 Section 6.4, Technical Processes (continued)

Section	*ISO 15288 Keywords*	*CMMI PAs*	*CMMI Practices*	*Confidence*	*Comment*
6.4.6.3b 4	Analyze information	PMC	SP 2.1	100	
		VER	SP 3.2	60	
6.4.7	**Transition Process**				
6.4.7.3	**Activities and tasks**				
6.4.7.3a	**Plan the transition**				
6.4.7.3a 1	Prepare strategy	PI	GP 2.2	30	ISO 15288 is much more detailed than CMMI
		PI	SP 3.4	30	
6.4.7.3a 2	Prepare site	PI	SP 3.4	30	
6.4.7.3b	**Perform the transition**				
6.4.7.3b 1	Deliver system	PI	SP 3.4	30	
6.4.7.3b 2	Install system	PI	SP 3.4	30	
6.4.7.3b 3	Demonstrate installation	PI	SP 3.4	30	
6.4.7.3b 4	Activate system	PI	SP 3.4	30	
6.4.7.3b 5	Demonstrate system	VAL	SP 2.2	60	
		PI	SP 3.4	30	
6.4.7.3b 6	Demonstrate sustainable service			0	
6.4.7.3b 7	Analyze information			0	
6.4.8	**Validation Process**				
6.4.8.3	**Activities and tasks**				
6.4.8.3a	**Plan validation**				
6.4.8.3a 1	Define strategy	VAL	GP 2.2	100	
		VAL	SP 1.1	100	
		VAL	SP 1.2	30	

6.4.8.3a 2	Prepare validation plan	VAL	GP 2.2	100
		VAL	SP 1.3	100
6.4.8.3b	**Perform validation**			
6.4.8.3b 1	Ensure operators are ready	VAL	SP 1.2	100
6.4.8.3b 2	Conduct validation	VAL	SP 2.1	100
6.4.8.3b 3	Make data available	VAL	SP 2.2	100
6.4.8.3b 4	Isolate nonconformance	VAL	SP 2.1	100
		VAL	SP 2.2	100
6.4.8.3b 5	Analyze data	VAL	SP 2.2	100
6.4.9	**Operation Process**			
6.4.9.3	**Activities and tasks**			
6.4.9.3a	**Prepare for operation**			0
6.4.9.3b	**Perform operational activation and checkout**			0
6.4.9.3c	**Use system for operations**			0
6.4.9.3d	**Perform operational problem resolution**			0
6.4.9.3e	**Support the customer**			0
6.4.10	**Maintenance Process**			
6.4.10.3	**Activities and tasks**			
6.4.10.3a	**Plan maintenance**			0
6.4.10.3b	**Perform maintenance**			0
6.4.11	**Disposal Process**			
6.4.11.3	**Activities and tasks**			
6.4.11.3a	**Plan disposal**			0
6.4.11.3b	**Perform disposal**			0
6.4.11.3c	**Finalize the disposal**			0

Appendix F: ISO 12207:2008 to CMMI v1.2 Map

Table F.1 ISO 12207:2008 Section 6.1, Agreement Processes

Section	*ISO 12207 Keywords*	*CMMI PAs*	*CMMI Practices*	*Confidence*	*Comment*
6	**System Life Cycle Processes**				
6.1	**Agreement Processes**				
6.1.1	**Acquisition Process**				
6.1.1.3	**Activities and tasks**				
6.1.1.3.1	**Acquisition preparation**				
6.1.1.3.1.1	Describe need for system			0	
6.1.1.3.1.2	Define and analyze system requirements	RD	SP 1.1	100	
		RD	SP 1.2	100	
		RD	SP 3.1	100	
		RD	SP 3.2	100	
6.1.1.3.1.3	Define and analyze software requirements	RD	SP 1.1	100	
		RD	SP 1.2	100	
		RD	SP 3.1	100	
		RD	SP 3.2	100	
6.1.1.3.1.4	Approve supplier's analysis			0	
6.1.1.3.1.5	Perform technical processes				See referenced section 6.4 requirements
6.1.1.3.1.6	Consider acquisition options	TS	SP 1.1	100	
		TS	SP 2.4	100	
6.1.1.3.1.7	Off-the-shelf software conditions	SAM	SP 1.2	30	
		SAM	SP 1.3	100	
		TS	SP 1.1	100	
6.1.1.3.1.8	Prepare acquisition plan	SAM	GP 2.2	60	

6.1.1.3.1.9	Acceptance criteria	SAM	SP 2.4	100	
6.1.1.3.1.10	Document acquisition requirements	SAM	SP 1.2	60	
6.1.1.3.1.11	Determine appropriate processes			0	
6.1.1.3.1.12	Define review milestones	SAM	SP 1.3	100	
6.1.1.3.1.13	Give requirements to acquisition organization	SAM	GP 2.2	60	CMMI elaboration notes use of groups outside of project
6.1.1.3.2	**Acquisition advertisement**				
6.1.1.3.2.1	Communicate request	SAM	SP 1.2	60	
6.1.1.3.3	**Supplier selection**				
6.1.1.3.3.1	Selection procedure	SAM	GP 2.2	100	
		SAM	SP 1.2	100	
6.1.1.3.3.2	Select supplier	SAM	SP 1.2	100	
6.1.1.3.4	**Contract agreement**				
6.1.1.3.4.1	Involve others in tailoring	SAM	GP 2.7	30	
6.1.1.3.4.2	Prepare contract	SAM	SP 1.3	100	
6.1.1.3.4.3	Control changes	SAM	SP 1.3	100	
6.1.1.3.5	**Agreement monitoring**				
6.1.1.3.5.1	Monitor activities				See referenced section 7 requirements
6.1.1.3.5.2	Provide information	SAM	SP 1.3	100	
		SAM	SP 2.1	100	
6.1.1.3.6	**Acquirer acceptance**				
6.1.1.3.6.1	Prepare for acceptance	SAM	SP 2.4	100	
6.1.1.3.6.2	Conduct acceptance testing	SAM	SP 2.4	100	
6.1.1.3.6.3	CM responsibility	SAM	SP 2.5	100	

(continued)

Table F.1 ISO 12207:2008 Section 6.1, Agreement Processes (continued)

Section	*ISO 12207 Keywords*	*CMMI PAs*	*CMMI Practices*	*Confidence*	*Comment*
6.1.1.3.7	**Closure**				
6.1.1.3.7.1	Make payment			0	
6.1.2	**Supply Process**				
6.1.2.3	**Activities and tasks**				
6.1.2.3.1	**Opportunity identification**				
6.1.2.3.1.1	Determine acquirer identity			0	Marketing activities not addressed in CMMI
6.1.2.3.2	**Supplier tendering**				
6.1.2.3.2.1	Review Request for Proposal			0	
6.1.2.3.2.2	Bid decision			0	
6.1.2.3.2.3	Prepare response			0	Proposal activities not addressed in CMMI
6.1.2.3.3	**Contract agreement**				
6.1.2.3.3.1	Negotiate	PP	SP 3.3	60	Contractual aspects of agreement not explicitly addressed in CMMI
		REQM	SP 1.2	60	
6.1.2.3.3.2	Contract modification			0	
6.1.2.3.4	**Contract execution**				
6.1.2.3.4.1	Review acquisition requirements			0	
6.1.2.3.4.2	Define life cycle	IPM	SP 1.1	100	
		PP	SP 1.3	100	
6.1.2.3.4.3	Establish requirements for plans	OPD	SP 1.1	100	
6.1.2.3.4.4	Consider development options	TS	SP 1.1	100	
		TS	SP 2.4	100	

6.1.2.3.4.5	Develop project management plans	IPM	SP 1.4	100	
		PP	SP 2.7	100	
6.1.2.3.4.6	Implement project management plans	IPM	SP 1.5	100	
		PMC	All SPs	100	
6.1.2.3.4.7	Develop, operate, and maintain				See referenced section 6 requirements
6.1.2.3.4.8	Monitor and control progress	IPM	SP 1.5	100	
		PMC	All SPs	100	
6.1.2.3.4.9	Manage subcontractors	SAM	SP 1.3	30	
		SAM	SP 2.1	100	
		SAM	SP 2.2	100	
		SAM	SP 2.3	100	
6.1.2.3.4.10	IV&V interface	PP	SP 2.6	30	
		PMC	SP 1.5	30	
6.1.2.3.4.11	Other party interfaces	PP	SP 2.6	100	
		PMC	SP 1.5	100	
6.1.2.3.4.12	Coordinate with acquirer	PP	SP 2.6	30	
		PMC	SP 1.5	30	
		PMC	SP 1.7	60	
6.1.2.3.4.13	Conduct meetings and reviews				See referenced section 7 requirements
6.1.2.3.4.14	Perform verification and validation				See referenced section 7 requirements
6.1.2.3.4.15	Make reports available			0	
6.1.2.3.4.16	Provide access to facilities			0	

(continued)

Table F.1 ISO 12207:2008 Section 6.1, Agreement Processes (continued)

Section	*ISO 12207 Keywords*	*CMMI PAs*	*CMMI Practices*	*Confidence*	*Comment*
6.1.2.3.4.17	Perform quality assurance				See referenced section 7.2.3 requirements
6.1.2.3.5	**Product/service delivery and support**				
6.1.2.3.5.1	Deliver product	PI	SP 3.4	100	
6.1.2.3.5.2	Provide assistance			0	
6.1.2.3.6	**Closure**				
6.1.2.3.6.1	Accept payment			0	
6.1.2.3.6.2	Transfer responsibility	PI	SP 3.4	30	

Table F.2 ISO 12207:2008 Section 6.2, Organizational Project-Enabling Processes

Section	*ISO 12207 Keywords*	*CMMI PAs*	*CMMI Practices*	*Confidence*	*Comment*
6.2	**Organizational Project-Enabling Processes**				
6.2.1	**Life Cycle Model Management Process**				
6.2.1.3	**Activities and tasks**				
6.2.1.3.1	**Process establishment**				
6.2.1.3.1.1	Establish organizational processes	OPD	GP 2.2	100	
		OPD	SP 1.1	100	
		OPD	SP 1.2	100	
		OPF	SP 3.4	100	
6.2.1.3.2	**Process assessment**				
6.2.1.3.2.1	Develop assessment procedure	OPF	SP 1.2	100	
6.2.1.3.2.2	Review processes	All PAs	GP 2.9	100	
6.2.1.3.3	**Process improvement**				
6.2.1.3.3.1	Improve processes	OPF	SP 1.3	100	
		OPF	SP 2.1	100	
		OPF	SP 2.2	100	
6.2.1.3.3.2	Collect historical data	OPF	SP 3.3	100	
		OPF	SP 3.4	100	
6.2.1.3.3.3	Collect quality cost data	OPD	SP 1.4	100	
		OPF	SP 3.4	100	
		All PAs	GP 2.8	60	
		All PAs	GP 3.2	100	

(continued)

Table F.2 ISO 12207:2008 Section 6.2, Organizational Project-Enabling Processes (continued)

Section	*ISO 12207 Keywords*	*CMMI PAs*	*CMMI Practices*	*Confidence*	*Comment*
6.2.2	**Infrastructure Management Process**				
6.2.2.3	**Activities and tasks**				
6.2.2.3.1	**Process implementation**				
6.2.2.3.1.1	Define infrastructure requirements	IPM	SP 1.3	100	
		OPD	SP 1.4	100	
		OPD	SP 1.5	100	
		OPD	SP 1.6	100	
6.2.2.3.1.2	Plan establishment of infrastructure	IPM	GP 2.2	100	
		OPD	GP 2.2	100	
6.2.2.3.2	**Establishment of the infrastructure**				
6.2.2.3.2.1	Plan infrastructure configuration	IPM	SP 1.3	100	
		OPD	GP 2.2	100	
6.2.2.3.2.2	Install infrastructure	IPM	SP 1.3	60	
		OPD	SP 1.4	60	
		OPD	SP 1.5	60	
		OPD	SP 1.6	60	
6.2.2.3.3	**Maintenance of the infrastructure**				
6.2.2.3.3.1	Maintain infrastructure	IPM	SP 1.3	100	
		OPD	SP 1.6	100	
6.2.3	**Project Portfolio Management Process**				
6.2.3.3	**Activities and tasks**				
6.2.3.3.1	**Project initiation**				
6.2.3.3.1.1	Identify business opportunities			0	

6.2.3.3.1.2	Define accountability	IPM	GP 2.4	100
		PMC	GP 2.4	100
		PP	GP 2.4	100
		PP	SP 2.7	100
6.2.3.3.1.3	Identify outcomes	PP	SP 1.1	60
6.2.3.3.1.4	Allocate resources	All PAs	GP 2.3	100
6.2.3.3.1.5	Identify project interfaces	IPM	SP 3.5	30
		OPD	SP 1.6	30
		PP	SP 2.6	60
6.2.3.3.1.6	Specify reporting requirements	PP	SP 2.6	30
		PP	SP 2.7	60
		PP	GP 2.4	100
6.2.3.3.1.7	Authorize project			0
6.2.3.3.2	**Portfolio evaluation**			
6.2.3.3.2.1	Evaluate projects	IPM	GP 2.10	100
		PMC	GP 2.10	100
		PPQA	GP 2.10	100
		PP	GP 2.10	100
6.2.3.3.2.2	Continue projects	IPM	GP 2.10	60
		PMC	GP 2.10	60
		PP	GP 2.10	60
6.2.3.3.3	**Project closure**			
6.2.3.3.3.1	Cancel projects	PP	GP 2.10	60
		PMC	GP 2.10	60
		IPM	GP 2.10	60
		RSKM	GP 2.10	60

(continued)

Table F.2 ISO 12207:2008 Section 6.2, Organizational Project-Enabling Processes (continued)

Section	*ISO 12207 Keywords*	*CMMI PAs*	*CMMI Practices*	*Confidence*	*Comment*
6.2.3.3.3.2	Close projects			0	
6.2.4	**Human Resource Management Process**				
6.2.4.3	**Activities and tasks**				
6.2.4.3.1	**Skill identification**				
6.2.4.3.1.1	Review project requirements	OT	SP 1.1	100	
		PP	SP 2.5	100	
6.2.4.3.1.2	Determine training type	OT	SP 1.1	100	
		PP	SP 2.5	100	
6.2.4.3.2	**Skill development**				
6.2.4.3.2.1	Develop training plan	OT	SP 1.3	100	
		PP	SP 2.5	100	
6.2.4.3.2.2	Develop training manuals	OT	SP 1.4	100	
6.2.4.3.2.3	Implement training plan	OT	SP 2.1	100	
6.2.4.3.3	**Skill acquisition and provision**				
6.2.4.3.3.1	Recruit staff			0	
6.2.4.3.3.2	Define evaluation criteria			0	
6.2.4.3.3.3	Evaluate staff performance			0	
6.2.4.3.3.4	Provide feedback to staff			0	
6.2.4.3.3.5	Maintain records of performance	OT	SP 2.2	100	
		PMC	GP 2.6	60	
6.2.4.3.3.6	Define teams	IPM	SP 3.2	100	
		OPD	SP 2.2	100	
6.2.4.3.3.7	Empower teams	IPM	SP 3.2	100	

		OPD	SP 2.1	100	
6.2.4.3.3.8	Ensure right mix of personnel	OT	SP 1.1	100	
		PP	SP 2.5	100	
6.2.4.3.4	**Knowledge management**				
6.2.4.3.4.1	Plan knowledge management			0	
6.2.4.3.4.2	Establish experts network			0	
6.2.4.3.4.3	Establish information exchange mechanism	OPD	SP 1.5	60	
6.2.4.3.4.4	Configuration manage assets				See referenced section 6.4.2 requirements
6.2.4.3.4.5	Maintain information	OPD	SP 1.5	60	
6.2.5	**Quality Management Process**				
6.2.5.3	**Activities and tasks**				
6.2.5.3.1	**Quality management**				
6.2.5.3.1.1	Establish policies and procedures	OPF	GP 2.1	100	QMS is equivalent to OSSP
		OPF	GP 2.2	100	
		OPD	GP 2.1	100	
		OPD	GP 2.2	100	
		OPD	SP 1.1	100	
6.2.5.3.1.2	Establish objectives	MA	SP 1.1	60	Must specify "customer satisfaction"
		OPF	SP 1.1	60	
		OPP	SP 1.3	60	
		QPM	SP 1.1	60	
6.2.5.3.1.3	Define responsibilities	MA	GP 2.4	100	
		OPF	GP 2.4	100	
		OPP	GP 2.4	100	
		PPQA	GP 2.4	100	

(continued)

Table F.2 ISO 12207:2008 Section 6.2, Organizational Project-Enabling Processes (continued)

Section	*ISO 12207 Keywords*	*CMMI PAs*	*CMMI Practices*	*Confidence*	*Comment*
		QPM	GP 2.4	100	
		OPD	GP 2.4	100	
6.2.5.3.1.4	Assess satisfaction	MA	SP 2.2	30	Weak and not explicit in CMMI
		MA	SP 2.4	30	
6.2.5.3.1.5	Review plans	All PAs	GP 2.9	60	
		All PAs	GP 2.10	60	
		PP	SP 3.1	100	
		IPM	SP 1.4	100	
6.2.5.3.1.6	Monitor status	All PAs	GP 2.8	30	
		OID	SP 2.3	100	
		OPF	SP 2.2	30	
		QPM	SP 2.3	100	
6.2.5.3.2	**Quality management corrective action**				
6.2.5.3.2.1	Plan corrective action	PPQA	SP 2.1	60	Planning is not explicit
6.2.5.3.2.2	Implement actions	PPQA	SP 2.2	100	Organizational communication not addressed

Table F.3 ISO 12207:2008 Section 6.3, Project Processes

Section	*ISO 12207 Keywords*	*CMMI PAs*	*CMMI Practices*	*Confidence*	*Comment*
6.3	**Project Processes**				
6.3.1	**Project Planning Process**				
6.3.1.3	**Activities and tasks**				
6.3.1.3.1	**Project initiation**				
6.3.1.3.1.1	Establish requirements for project	PP	SP 1.1	100	
6.3.1.3.1.2	Establish feasibility	PP	SP 3.1	100	
6.3.1.3.1.3	Modify project requirements	PP	SP 3.2	100	
6.3.1.3.2	**Project planning**				
6.3.1.3.2.1	Prepare plans	All PAs	GP 2.2	100	
		PP	All SPs	100	
6.3.1.3.3	**Project activation**				
6.3.1.3.3.1	Authorize project			0	
6.3.1.3.3.2	Request resources	PP	SP 3.2	100	
6.3.1.3.3.3	Initiate implementation	IPM	SP 1.5	60	"Initiation" not specifically addressed
6.3.2	**Project Assessment and Control Process**				
6.3.2.3	**Activities and tasks**				
6.3.2.3.1	**Project monitoring**				
6.3.2.3.1.1	Monitor execution	PMC	SP 1.1	100	
		PMC	SP 1.2	100	
		PMC	SP 1.3	100	
		PMC	SP 1.4	100	
		PMC	SP 1.5	100	

(continued)

Table F.3 ISO 12207:2008 Section 6.3, Project Processes (continued)

Section	*ISO 12207 Keywords*	*CMMI PAs*	*CMMI Practices*	*Confidence*	*Comment*
6.3.2.3.2	**Project control**				
6.3.2.3.2.1	Investigate problems	PMC	SP 2.1	100	
		PMC	SP 2.2	100	
		PMC	SP 2.3	100	
6.3.2.3.2.2	Report project progress	All PAs	GP 2.10	100	
		PMC	SP 1.6	100	
		PMC	SP 1.7	100	
6.3.2.3.3	**Project assessment**				
6.3.2.3.3.1	Evaluate products	All PAs	GP 2.9	100	
		PPQA	SP 1.2	100	
6.3.2.3.3.2	Assess evaluation results	All PAs	GP 2.9	100	
6.3.2.3.4	**Project closure**				
6.3.2.3.4.1	Determine project completion			0	Nothing about "closure" in CMMI
6.3.2.3.4.2	Archive results			0	Nothing about "archiving" in CMMI
6.3.3	**Decision Management Process**				
6.3.3.3	**Activities and tasks**				
6.3.3.3.1	**Decision planning**				
6.3.3.3.1.1	Define strategy	DAR	GP 2.2	100	
		DAR	SP 1.1	100	
6.3.3.3.1.2	Involve relevant parties	DAR	GP 2.7	100	
6.3.3.3.1.3	Identify need for decision	DAR	SP 1.1	100	
		DAR	SP 1.2	100	
		DAR	SP 1.3	100	

6.3.3.3.2	**Decision analysis**			
6.3.3.3.2.1	Select strategy	DAR	SP 1.4	100
6.3.3.3.2.2	Evaluate alternatives	DAR	SP 1.5	100
6.3.3.3.3	**Decision tracking**			
6.3.3.3.3.1	Record decisions	DAR	SP 1.6	100
6.3.3.3.3.2	Maintain records	DAR	GP 2.6	100
		DAR	SP 1.6	100
6.3.4	**Risk Management Process**			
6.3.4.3	**Activities and tasks**			
6.3.4.3.1	**Risk management planning**			
6.3.4.3.1.1	Define policies	RSKM	GP 2.1	100
6.3.4.3.1.2	Document risk process	RSKM	GP 2.2	100
		RSKM	GP 3.1	100
6.3.4.3.1.3	Identify responsible parties	RSKM	GP 2.4	100
		RSKM	GP 2.7	100
6.3.4.3.1.4	Provide resources	RSKM	GP 2.3	100
6.3.4.3.1.5	Define risk improvement process	RSKM	GP 2.8	30
		RSKM	GP 3.2	60
6.3.4.3.2	**Risk profile management**			
6.3.4.3.2.1	Document risk context	RSKM	SP 1.1	100
6.3.4.3.2.2	Define thresholds	RSKM	SP 1.2	100
6.3.4.3.2.3	Establish risk profile	RSKM	SP 2.2	100
6.3.4.3.2.4	Communicate risk profile	RSKM	GP 2.7	100
6.3.4.3.3	**Risk analysis**			
6.3.4.3.3.1	Identify risks	PP	SP 2.2	100
		RSKM	SP 2.1	100

(continued)

Table F.3 ISO 12207:2008 Section 6.3, Project Processes (continued)

Section	*ISO 12207 Keywords*	*CMMI PAs*	*CMMI Practices*	*Confidence*	*Comment*
6.3.4.3.3.2	Estimate probability	RSKM	SP 2.2	100	
6.3.4.3.3.3	Evaluate risks	RSKM	SP 2.2	100	
6.3.4.3.3.4	Define risk treatment strategy	RSKM	SP 3.1	100	
6.3.4.3.4	**Risk treatment**				
6.3.4.3.4.1	Recommend risk treatment	RSKM	SP 3.1	100	
6.3.4.3.4.2	Implement risk treatment	RSKM	SP 3.2	100	
6.3.4.3.4.3	Monitor risks	RSKM	SP 3.2	100	
6.3.4.3.4.4	Manage IAW section 6.3.2 or 15288				See referenced requirements
6.3.4.3.5	**Risk monitoring**				
6.3.4.3.5.1	Monitor risks and context	PMC	SP 1.3	100	
		RSKM	SP 3.2	100	
6.3.4.3.5.2	Monitor measures	RSKM	GP 2.8	100	
6.3.4.3.5.3	Monitor new risks	PP	SP 2.1	100	
		RSKM	SP 2.1	100	
6.3.4.3.6	**Risk management process evaluation**				
6.3.4.3.6.1	Collect improvement information	RSKM	GP 3.2	100	
6.3.4.3.6.2	Review risk management process	RSKM	GP 2.8	30	Effectiveness and efficiency not explicit
6.3.4.3.6.3	Review risk information	OPF	SP 3.4	100	This should be specifically addressed in reviews
		RSKM	GP 3.2	60	
		RSKM	GP 2.10	60	
6.3.5	**Configuration Management Process**				
6.3.5.3	**Activities and tasks**				

6.3.5.3.1	**Configuration management planning**				
6.3.5.3.1.1	Define CM strategy	CM	GP 2.2	100	
		CM	SP 1.2	100	
6.3.5.3.1.2	Identify items	CM	SP 1.1	100	
6.3.5.3.2	**Configuration management execution**				
6.3.5.3.2.1	Maintain configuration information	CM	SP 1.3	100	
		CM	SP 2.1	100	
		CM	SP 3.1	100	
6.3.5.3.2.2	Maintain baselines	CM	SP 2.1	100	
		CM	SP 2.2	100	
		CM	SP 3.1	100	
		CM	SP 3.2	100	
6.3.6	**Information Management Process**				
6.3.6.3	**Activities and tasks**				
6.3.6.3.1	**Information management planning**				
6.3.6.3.1.1	Define items to be managed	PP	SP 2.3	100	
6.3.6.3.1.2	Designate responsibilities	PP	SP 2.3	100	
6.3.6.3.1.3	Define rights	PP	SP 2.3	100	
6.3.6.3.1.4	Define representation	PP	SP 2.3	100	
6.3.6.3.1.5	Define maintenance	PP	SP 2.3	100	
6.3.6.3.2	**Information management execution**				
6.3.6.3.2.1	Obtain information items	PMC	SP 1.4	30	Data collection activities are not explicitly identified in CMMI
6.3.6.3.2.2	Maintain information items	PMC	SP 1.4	100	
6.3.6.3.2.3	Retrieve information	PMC	SP 1.4	30	
6.3.6.3.2.4	Provide documentation	PMC	SP 1.4	30	

(continued)

Table F.3 ISO 12207:2008 Section 6.3, Project Processes (continued)

Section	*ISO 12207 Keywords*	*CMMI PAs*	*CMMI Practices*	*Confidence*	*Comment*
6.3.6.3.2.5	Archive information	PMC	SP 1.4	30	
6.3.6.3.2.6	Dispose of information	PMC	SP 1.4	30	
6.3.7	**Measurement Process**				
6.3.7.3	**Activities and tasks**				
6.3.7.3.1	**Measurement planning**				
6.3.7.3.1.1	Describe relevant characteristics	MA	SP 1.1	100	
6.3.7.3.1.2	Identify information needs	MA	SP 1.1	100	
6.3.7.3.1.3	Select measures	MA	SP 1.2	100	
6.3.7.3.1.4	Define procedures	MA	SP 1.3	100	
6.3.7.3.1.5	Define evaluation criteria	MA	SP 1.4	100	
6.3.7.3.1.6	Provide resources	MA	GP 2.3	100	
6.3.7.3.1.7	Deploy technologies	MA	GP 2.3	100	
6.3.7.3.2	**Measurement performance**				
6.3.7.3.2.1	Integrate procedures	MA	GP 2.2	60	Integration is not explicit in CMMI
6.3.7.3.2.2	Collect data	MA	SP 2.1	100	
		MA	SP 2.3	100	
6.3.7.3.2.3	Analyze data	MA	SP 2.2	100	
6.3.7.3.2.4	Document results	MA	SP 2.4	100	
6.3.7.3.3	**Measurement evaluation**				
6.3.7.3.3.1	Evaluate products and process	MA	GP 2.9	100	
6.3.7.3.3.2	Identify improvements	MA	GP 3.2	100	

Table F.4 ISO 12207:2008 Section 6.4, Technical Processes

Section	*ISO 12207 Keywords*	*CMMI PAs*	*CMMI Practices*	*Confidence*	*Comment*
6.4	**Technical Processes**				
6.4.1	**Stakeholder Requirements Definition Process**				
6.4.1.3	**Activities and tasks**				
6.4.1.3.1	**Stakeholder identification**				
6.4.1.3.1.1	Identify stakeholders	All PAs	GP 2.7	100	
		PP	SP 2.6	100	
6.4.1.3.2	**Requirements identification**				
6.4.1.3.2.1	Elicit requirements	RD	SP 1.1	100	
6.4.1.3.2.2	Define constraints	RD	SP 1.1	100	
		RD	SP 1.2	100	
6.4.1.3.2.3	Define operational scenarios	RD	SP 3.1	100	
6.4.1.3.2.4	Identify interactions	RD	SP 3.1	100	
6.4.1.3.2.5	Specify environmental requirements	RD	SP 1.1	60	
		RD	SP 1.2	60	
		RD	SP 2.1	60	
		RD	SP 2.3	100	
		RD	SP 3.2	60	
6.4.1.3.3	**Requirements evaluation**				
6.4.1.3.3.1	Analyze requirements	RD	SP 3.3	100	
6.4.1.3.4	**Requirements agreement**				
6.4.1.3.4.1	Resolve problems	RD	SP 3.3	100	
		RD	SP 3.4	30	

(continued)

Table F.4 ISO 12207:2008 Section 6.4, Technical Processes (continued)

Section	*ISO 12207 Keywords*	*CMMI PAs*	*CMMI Practices*	*Confidence*	*Comment*
6.4.1.3.4.2	Feedback requirements	RD	SP 3.5	100	
6.4.1.3.4.3	Confirm requirements understanding	RD	SP 3.5	100	
6.4.1.3.5	**Requirement recording**				
6.4.1.3.5.1	Record requirements	RD	SP 1.2	100	
		RD	SP 2.2	100	
		REQM	GP 2.6	100	
		REQM	SP 1.3	100	
6.4.1.3.5.2	Maintain traceability	REQM	SP 1.4	100	
6.4.2	**System Requirements Analysis Process**				
6.4.2.3	**Activities and tasks**				
6.4.2.3.1	**Requirements specification**				
6.4.2.3.1.1	Analyze intended use	RD	SP 3.1	100	
		RD	SP 3.2	100	
6.4.2.3.2	**Requirements evaluation**				
6.4.2.3.2.1	Evaluate system requirements	RD	SP 2.1	100	
		RD	SP 3.3	100	
		RD	SP 3.4	60	
		REQM	SP 1.4	100	
		VER	SP 2.1	100	
		VER	SP 2.2	100	
		VER	SP 2.3	100	
6.4.3	**System Architectural Design Process**				
6.4.3.3	**Activities and tasks**				

6.4.3.3.1	**Establishing architecture**				
6.4.3.3.1.1	Establish top level architecture	RD	SP 2.2	100	
		TS	SP 1.1	100	
		TS	SP 1.2	100	
		TS	SP 2.1	100	
6.4.3.3.2	**Architectural evaluation**				
6.4.3.3.2.1	Evaluate architecture	RD	SP 2.2	100	
		REQM	SP 1.4	100	
		TS	SP 1.1	100	
		TS	SP 1.2	100	
		VER	SP 2.1	100	
		VER	SP 2.2	100	
		VER	SP 2.3	100	
6.4.4	**Implementation Process**				
6.4.5	**System Integration Process**				
6.4.5.3	**Activities and tasks**				
6.4.5.3.1	**Integration**				
6.4.5.3.1.1	Integrate and test configuration items	PI	SP 3.2	100	
		PI	SP 3.3	100	
		VER	SP 3.1	100	
6.4.5.3.2	**Test readiness**				
6.4.5.3.2.1	Develop tests and ensure readiness for qualification testing	VER	SP 1.3	100	
		VER	SP 3.1	100	
		VER	SP 3.2	100	
6.4.5.3.2.2	Evaluate integrated system			0	ISO is more detailed than CMMI

(continued)

Table F.4 ISO 12207:2008 Section 6.4, Technical Processes (continued)

Section	*ISO 12207 Keywords*	*CMMI PAs*	*CMMI Practices*	*Confidence*	*Comment*
6.4.6	**System Qualification Testing Process**				
6.4.6.3	**Activities and tasks**				
6.4.6.3.1	**Qualification testing**				
6.4.6.3.1.1	Conduct qualification testing	VER	SP 3.1	100	
6.4.6.3.1.2	Evaluate system				
6.4.6.3.1.2a	Evaluate coverage	VER	SP 3.2	100	
6.4.6.3.1.2b	Evaluate results	VER	SP 3.2	100	
6.4.6.3.1.2c	Evaluate operation and maintenance			0	
6.4.6.3.1.3	Support audits IAW 7.2.7				See referenced section 7.2.7 requirements
6.4.6.3.1.4	Prepare product for installation	PI	SP 3.4	100	Also refer to section 7.2.4 and 7.2.5
6.4.7	**Software Installation Process**				
6.4.7.3	**Activities and tasks**				
6.4.7.3.1	**Software installation**				
6.4.7.3.1.1	Plan installation	PI	GP 2.2	30	PI plan must address installation
6.4.7.3.1.2	Install software produce	PI	SP 3.4	100	
6.4.8	**Software Acceptance Support Process**				
6.4.8.3	**Activities and tasks**				
6.4.8.3.1	**Software acceptance support**				

6.4.8.3.1.1	Support acceptance review	PI	SP 3.4	30	Also refer to section 7.2.6 and 7.2.7; ISO is more detailed than CMMI
		VAL	SP 2.1	60	
		VAL	SP 2.2	60	
6.4.8.3.1.2	Deliver software product	PI	SP 3.4	100	
6.4.8.3.1.3	Provide support			0	
6.4.9	**Software Operation Process**				
6.4.9.3	**Activities and tasks**				
6.4.9.3.1	**Preparation for operation**			0	
6.4.9.3.2	**Operation activation and checkout**			0	
6.4.9.3.3	**Operational use**			0	
6.4.9.3.4	**Customer support**			0	
6.4.9.3.5	**Operation problem resolution**			0	
6.4.10	**Software Maintenance Process**				
6.4.10.3	**Activities and tasks**				
6.4.10.3.1	**Process implementation**			0	
6.4.10.3.2	**Problem and modification analysis**			0	
6.4.10.3.3	**Modification implementation**			0	
6.4.10.3.4	**Maintenance review/acceptance**			0	
6.4.10.3.5	**Migration**			0	
6.4.11	**Software Disposal Process**				
6.4.11.3	**Activities and tasks**				
6.4.11.3.1	**Software disposal planning**			0	
6.4.11.3.2	**Software disposal execution**			0	

Table F.5 ISO 12207:2008 Section 7.1, Software Life Cycle Processes

Section	*ISO 12207 Keywords*	*CMMI PAs*	*CMMI Practices*	*Confidence*	*Comment*
7	**Software Life Cycle Processes**				
7.1	**Software Implementation Processes**				
7.1.1	**Software Implementation Process**				
7.1.1.3	**Activities and tasks**				
7.1.1.3.1	**Software implementation strategy**				
7.1.1.3.1.1	Define life cycle model	IPM	SP 1.1	100	
		PP	SP 1.3	100	
7.1.1.3.1.2	Perform software support processes				See referenced section 7.2 requirements
7.1.1.3.1.3	Tailor standards	IPM	SP 1.1	100	
7.1.1.3.1.4	Develop software implementation plans	PI	GP 2.2	100	
		RD	GP 2.2	100	
		REQM	GP 2.2	100	
		TS	GP 2.2	100	
		VAL	GP 2.2	100	
		VER	GP 2.2	100	
7.1.1.3.1.5	Ensure independence of nondeliverable items			0	
7.1.2	**Software Requirements Analysis Process**				
7.1.2.3	**Activities and tasks**				
7.1.2.3.1	**Software requirements analysis**				
7.1.2.3.1.1	Document software requirements	RD	SP 2.1	60	Safety, security, operation, and maintenance are not explicitly addressed in CMMI
		RD	SP 2.2	60	
		RD	SP 2.3	60	

		RD	SP 3.1	100	
		RD	SP 3.2	100	
7.1.2.3.1.2	Evaluate software requirements	RD	SP 3.3	60	Operation and maintenance are not explicitly addressed in CMMI
		REQM	SP 1.4	100	
		VER	SP 2.1	100	
		VER	SP 2.2	100	
		VER	SP 2.3	100	
7.1.2.3.1.3	Conduct reviews				See referenced section 7.2.6 requirements
7.1.3	**Software Architectural Design Process**				
7.1.3.3	**Activities and tasks**				
7.1.3.3.1	**Software architectural design**				
7.1.3.3.1.1	Transform requirements into architecture	REQM	SP 1.4	100	
		TS	SP 2.1	100	
7.1.3.3.1.2	Develop top-level design	TS	SP 2.1	100	
7.1.3.3.1.3	Develop top-level database design	TS	SP 2.1	30	Database design not explicitly addressed
7.1.3.3.1.4	Develop preliminary user documents	TS	SP 3.2	100	
7.1.3.3.1.5	Define preliminary test requirements	PI	SP 1.3	100	
		VER	SP 1.3	100	
7.1.3.3.1.6	Evaluate architecture	TS	SP 2.1	60	Operation and maintenance feasibility not addressed in CMMI
		VER	SP 2.1	100	
		VER	SP 2.2	100	
		VER	SP 2.3	100	
7.1.3.3.1.7	Conduct reviews				See referenced section 7.2.6 requirements

(continued)

Table F.5 ISO 12207:2008 Section 7.1, Software Life Cycle Processes (continued)

Section	*ISO 12207 Keywords*	*CMMI PAs*	*CMMI Practices*	*Confidence*	*Comment*
7.1.4	**Software Detailed Design Process**				
7.1.4.3	**Activities and tasks**				
7.1.4.3.1	**Software detailed design**				
7.1.4.3.1.1	Develop detailed designs	TS	SP 2.1	100	
7.1.4.3.1.2	Develop detailed interface design	TS	SP 2.3	100	
7.1.4.3.1.3	Develop detailed database design	TS	SP 2.1	30	Database design not explicitly addressed
7.1.4.3.1.4	Update user documentation	TS	SP 3.2	100	
7.1.4.3.1.5	Define unit test requirements and schedule	VER	SP 1.1	100	
		VER	SP 1.3	100	
7.1.4.3.1.6	Update integration test requirements and schedule	PI	SP 1.3	100	
		VER	SP 1.3	100	
7.1.4.3.1.7	Evaluate detailed design	TS	SP 2.1	100	Operation and maintenance feasibility not addressed in CMMI
		VER	SP 3.1	100	
		VER	SP 3.2	100	
7.1.4.3.1.8	Conduct reviews				See referenced section 7.2.6 requirements
7.1.5	**Software Construction Process**				
7.1.5.3	**Activities and tasks**				
7.1.5.3.1	**Software construction**				
7.1.5.3.1.1	Develop software units and tests	TS	SP 3.1	100	

7.1.5.3.1.2	Test software units	TS	SP 3.1	100	
7.1.5.3.1.3	Update user documentation	TS	SP 3.2	100	
7.1.5.3.1.4	Update integration test requirements and schedule	PI	SP 1.1	100	
		PI	SP 1.3	100	
		VER	SP 1.3	100	
7.1.5.3.1.5	Evaluate code	TS	SP 3.1	100	Operation and maintenance feasibility not addressed in CMMI
		VER	SP 2.1	100	
		VER	SP 2.2	100	
		VER	SP 2.3	100	
7.1.6	**Software Integration Process**				
7.1.6.3	**Activities and tasks**				
7.1.6.3.1	**Software integration**				
7.1.6.3.1.1	Develop unit integration plan	PI	GP 2.2	100	
		PI	SP 1.1	100	
		PI	SP 1.3	100	
7.1.6.3.1.2	Integrate software units	PI	SP 3.2	100	
		PI	SP 3.3	100	
7.1.6.3.1.3	Update user documentation	TS	SP 3.2	100	
7.1.6.3.1.4	Develop qualification tests	PI	SP 1.3	30	
7.1.6.3.1.5	Evaluate integration	PI	GP 2.2	100	Operation and maintenance feasibility not addressed in CMMI
		PI	SP 3.3	60	
		REQM	SP 1.4	100	
		REQM	SP 1.5	100	
		VER	All SPs	100	

(continued)

Table F.5 ISO 12207:2008 Section 7.1, Software Life Cycle Processes (continued)

Section	*ISO 12207 Keywords*	*CMMI PAs*	*CMMI Practices*	*Confidence*	*Comment*
7.1.6.3.1.6	Conduct reviews				See referenced section 7.2.6 requirements
7.1.7	**Software Qualification Testing Process**				
7.1.7.3	**Activities and tasks**				
7.1.7.3.1	**Software qualification testing**				
7.1.7.3.1.1	Conduct qualification testing	REQM	SP 1.4	100	
		VER	SP 3.1	100	
7.1.7.3.1.2	Update user documentation	TS	SP 3.2	100	
7.1.7.3.1.3	Evaluate design, code, documentation	VER	SP 3.2	60	Operation and maintenance feasibility not addressed in CMMI
7.1.7.3.1.4	Support audits				See referenced section 7.2.7 requirements
7.1.7.3.1.5	Update and prepare software product	PMC	SP 2.1	60	Software qualification testing may support software validation and software verification
		PMC	SP 2.2	60	
		PMC	SP 2.3	60	

Table F.6 ISO 12207:2008 Section 7.2, Software Support Processes

Section	*ISO 12207 Keywords*	*CMMI PAs*	*CMMI Practices*	*Confidence*	*Comment*
7.2	**Software Support Processes**				
7.2.1	**Software Documentation Management Process**				
7.2.1.3	**Activities and tasks**				
7.2.1.3.1	**Process implementation**				
7.2.1.3.1.1	Develop documentation plan	All PAs	GP 2.2	60	
7.2.1.3.2	**Design and development**				
7.2.1.3.2.1	Use documentation standards	All PAs	GP 2.2	100	
		OPD	SP 1.1	100	
7.2.1.3.2.2	Confirm input data	TS	SP 2.2	60	TS practices alone do not address the breadth of the ISO 12207 requirement
		TS	SP 3.2	60	
7.2.1.3.2.3	Review against document standards	PPQA	SP 1.2	100	
		VER	SP 2.1	100	
		VER	SP 2.2	100	
		VER	SP 2.3	100	
7.2.1.3.3	**Production**				
7.2.1.3.3.1	Produce documents as planned	IPM	SP 1.5	30	Documentation not explicitly addressed
7.2.1.3.3.2	Control documents using CM process				See referenced section 7.2.2 requirements
7.2.1.3.4	**Maintenance**				
7.2.1.3.4.1	Perform document maintenance	CM	All SPs	100	See referenced section 6.4.10 and 7.2.2 requirements

(continued)

Table F.6 ISO 12207:2008 Section 7.2, Software Support Processes (Continued)

Section	*ISO 12207 Keywords*	*CMMI PAs*	*CMMI Practices*	*Confidence*	*Comment*
7.2.2	**Software Configuration Management Process**				
7.2.2.3	**Activities and tasks**				
7.2.2.3.1	**Process implementation**				
7.2.2.3.1.1	Develop configuration management plan	CM	GP 2.2	100	
7.2.2.3.2	**Configuration identification**				
7.2.2.3.2.1	Establish configuration identification scheme	CM	SP 1.1	100	
7.2.2.3.3	**Configuration control**				
7.2.2.3.3.1	Control changes	CM	SP 2.1	100	
		CM	SP 2.2	100	
7.2.2.3.4	**Configuration status accounting**				
7.2.2.3.4.1	Prepare configuration status reports	CM	SP 3.1	100	
7.2.2.3.5	**Configuration evaluation**				
7.2.2.3.5.1	Ensure functional and physical completeness	CM	SP 3.2	100	
7.2.2.3.6	**Release management and delivery**				
7.2.2.3.6.1	Control product release	CM	SP 1.3	100	
7.2.3	**Software Quality Assurance Process**				
7.2.3.3	**Activities and tasks**				
7.2.3.3.1	**Process implementation**				
7.2.3.3.1.1	Establish quality assurance process	PPQA	GP 2.2	100	
7.2.3.3.1.2	Coordinate QA with related processes				See referenced section 7.2.4, 7.2.5, 7.2.6, and 7.2.7 requirements

7.2.3.3.1.3	Plan quality assurance activities	PPQA	GP 2.2	100	See referenced section 7.2.4, 7.2.5, 7.2.6, 7.2.7, and 7.2.8 requirements
7.2.3.3.1.4	Execute quality assurance activities	PPQA	SP 1.1	100	See referenced section 7.2.8 requirements
		PPQA	SP 1.2	100	
		PPQA	SP 2.1	100	
		PPQA	SP 2.2	100	
7.2.3.3.1.5	Make QA records available	PPQA	SP 2.2	100	
7.2.3.3.1.6	Assure organizational freedom and authority	PPQA	SP 1.1	100	CMMI uses "objectively evaluate"
		PPQA	SP 1.2	100	
7.2.3.3.2	**Product assurance**				
7.2.3.3.2.1	Assure plans are documented	PPQA	SP 1.2	100	ISO specifically requires "plans"
7.2.3.3.2.2	Assure software complies with plans	PPQA	SP 1.2	100	ISO specifically requires "products adhering to plans"
7.2.3.3.2.3	Assure products satisfy contract	PPQA	SP 1.2	30	
		VER	SP 3.1	100	
		VER	SP 3.2	100	
7.2.3.3.3	**Process assurance**				
7.2.3.3.3.1	Assure processes comply with contract and plans	PPQA	SP 1.1	100	ISO specifically includes "plans"
7.2.3.3.3.2	Assure software engineering practices comply with contract	PPQA	SP 1.1	30	
7.2.3.3.3.3	Assure contract requirements are passed to subcontractors	SAM	GP 2.9	60	
7.2.3.3.3.4	Assure that acquirer and others are supported	IPM	GP 2.9	60	

(continued)

Table F.6 ISO 12207:2008 Section 7.2, Software Support Processes (Continued)

Section	*ISO 12207 Keywords*	*CMMI PAs*	*CMMI Practices*	*Confidence*	*Comment*
7.2.3.3.3.5	Assure measurements follow standards	PPQA	SP 1.1	60	CMMI does not explicitly address assurance of measurements
		PPQA	SP 1.2	60	
		MA	GP 2.9	100	
7.2.3.3.3.6	Assure that staff has needed skills	PPQA	SP 1.1	60	CMMI does not explicitly address assurance of training
		PPQA	SP 1.2	60	
		PMC	GP 2.9	60	
7.2.3.3.4	**Assurance of quality systems**				
7.2.3.3.4.1	Additional activities IAW ISO 9001	OPF	SP 1.2	100	
7.2.4	**Software Verification Process**				
7.2.4.3	**Activities and tasks**				
7.2.4.3.1	**Process implementation**				
7.2.4.3.1.1	Determine need for verification effort	VER	SP 1.1	60	
7.2.4.3.1.2	Establish verification process	VER	GP 2.2	100	
7.2.4.3.1.3	Select independent verification organization			0	
7.2.4.3.1.4	Determine activities and products for verification	VER	SP 1.1	100	
7.2.4.3.1.5	Develop verification plan	VER	GP 2.2	100	
		VER	GP 2.3	100	
		VER	GP 2.4	100	
7.2.4.3.1.6	Implement verification plan	VER	SP 2.1	100	See referenced section 7.2.8 requirements
		VER	SP 2.2	100	
		VER	SP 2.3	100	

		VER	SP 3.1	100	
		VER	SP 3.2	100	
7.2.4.3.2	**Verification**				
7.2.4.3.2.1	Requirements verification	RD	SP 3.3	100	
7.2.4.3.2.2	Design verification	REQM	SP 1.4	100	Safety and security are not explicitly addressed
		TS	SP 2.1	60	
7.2.4.3.2.3	Code verification	REQM	SP 1.4	100	Safety and security are not explicitly addressed
		TS	SP 3.1	60	
7.2.4.3.2.4	Integration verification	PI	SP 3.3	100	
7.2.4.3.2.5	Documentation verification	TS	GP 2.6	100	
		TS	SP 2.2	100	
		TS	SP 3.2	100	
7.2.5	**Software Validation Process**				
7.2.5.3	**Activities and tasks**				
7.2.5.3.1	**Process implementation**				
7.2.5.3.1.1	Determine need for validation effort	VAL	SP 1.1	60	
7.2.5.3.1.2	Establish validation process	VAL	GP 2.2	100	
7.2.5.3.1.3	Select independent validation organization			0	
7.2.5.3.1.4	Develop validation plan	VAL	GP 2.2	100	
		VAL	GP 2.3	100	
		VAL	GP 2.4	100	
7.2.5.3.1.5	Implement validation plan	VAL	SP 2.1	100	See referenced section 7.2.8 requirements
		VAL	SP 2.2	100	

(continued)

Table F.6 ISO 12207:2008 Section 7.2, Software Support Processes (Continued)

Section	*ISO 12207 Keywords*	*CMMI PAs*	*CMMI Practices*	*Confidence*	*Comment*
7.2.5.3.2	**Validation**				
7.2.5.3.2.1	Prepare validation requirements and specifications	VAL	SP 1.1	100	
		VAL	SP 1.2	100	
		VAL	SP 1.3	100	
7.2.5.3.2.2	Ensure validation test reflects intended use	VAL	GP 2.9	60	
7.2.5.3.2.3	Conduct test	VAL	SP 2.1	100	
7.2.5.3.2.4	Validate software product	VAL	SP 2.2	100	
7.2.5.3.2.5	Test in target environment	VAL	SP 2.1	100	
7.2.6	**Software Review Process**				
7.2.6.3	**Activities and tasks**				
7.2.6.3.1	**Process implementation**				
7.2.6.3.1.1	Conduct periodic reviews	PMC	SP 1.7	100	
7.2.6.3.1.2	Agree on resources	PMC	GP 2.3	60	
		PP	SP 2.4	100	
7.2.6.3.1.3	Agree on agenda and scope	PMC	GP 2.7	100	
7.2.6.3.1.4	Record problems	PMC	SP 2.1	100	See referenced section 7.2.8 requirements
7.2.6.3.1.5	Document review results	PMC	SP 1.7	100	
7.2.6.3.1.6	Agree on outcome and actions	PMC	SP 1.7	60	

7.2.6.3.2	**Project Management Reviews**				
7.2.6.3.2.1	Evaluate project status	PMC	GP 2.10	100	
		PMC	SP 1.3	100	
		PMC	SP 1.6	100	
		PMC	SP 1.7	100	
7.2.6.3.3	**Technical Reviews**				
7.2.6.3.3.1	Conduct technical reviews	PMC	SP 1.7	100	See referenced section 7.2.2 requirements
		VER	SP 2.1	100	
		VER	SP 2.2	100	
		VER	SP 2.3	100	
		VER	SP 3.1	100	
		VER	SP 3.2	100	
7.2.7	**Software Audit Process**				
7.2.7.3	**Activities and tasks**				
7.2.7.3.1	**Process implementation**				
7.2.7.3.1.1	Hold audits at milestones	PPQA	GP 2.2	100	
7.2.7.3.1.2	Auditors not responsible for products	PPQA	SP 1.1	60	CMMI requires "objective evaluation"
		PPQA	SP 1.2	60	
7.2.7.3.1.3	Resources agreed upon	PPQA	GP 2.3	100	
7.2.7.3.1.4	Audit scope agreed	PPQA	GP 2.7	100	
7.2.7.3.1.5	Problems recorded	PPQA	SP 2.1	100	See referenced section 7.2.8 requirements
7.2.7.3.1.6	Audit results documented	PPQA	SP 2.1	100	
7.2.7.3.1.7	Audit outcome agreed	PPQA	SP 2.1	100	

(continued)

Table F.6 ISO 12207:2008 Section 7.2, Software Support Processes (Continued)

Section	*ISO 12207 Keywords*	*CMMI PAs*	*CMMI Practices*	*Confidence*	*Comment*
7.2.7.3.2	**Software audit**				
7.2.7.3.2.1	Conduct audits	PPQA	SP 1.1	100	
		PPQA	SP 1.2	100	
7.2.8	**Software Problem Resolution Process**				
7.2.8.3	**Activities and tasks**				
7.2.8.3.1	**Process implementation**				
7.2.8.3.1.1	Establish problem resolution process	PMC	GP 2.2	100	
7.2.8.3.2	**Problem resolution**				
7.2.8.3.2.1	Report and resolve problems	PMC	SP 2.1	100	
		PMC	SP 2.2	100	
		PMC	SP 2.3	100	

Table F.7 ISO 12207:2008 Section 7.3, Software Reuse Processes

Section	*ISO 12207 Keywords*	*CMMI PAs*	*CMMI Practices*	*Confidence*	*Comment*
7.3	**Software Reuse Processes**				
7.3.1	**Domain Engineering Process**				
7.3.1.3	**Activities and tasks**				
7.3.1.3.1	**Process implementation**			0	
7.3.1.3.2	**Domain analysis**			0	
7.3.1.3.3	**Domain design**			0	
7.3.1.3.4	**Asset provision**			0	
7.3.1.3.5	**Asset maintenance**			0	
7.3.2	**Reuse Asset Management Process**				
7.3.2.3	**Activities and tasks**				
7.3.2.3.1	**Process implementation**			0	
7.3.2.3.2	**Asset storage and retrieval definition**			0	
7.3.2.3.3	**Asset management and control**			0	
7.3.3	**Reuse Program Management Process**				
7.3.3.3	**Activities and tasks**				
7.3.3.3.1	**Initiation**			0	
7.3.3.3.2	**Domain identification**			0	
7.3.3.3.3	**Reuse assessment**			0	
7.3.3.3.4	**Planning**			0	
7.3.3.3.5	**Execution and control**			0	
7.3.3.3.6	**Review and evaluation**			0	

Appendix G: ISO 20000:2005 to CMMI v1.2 Map

Table G.1 ISO 20000:2005 Section 3, Requirements for a Management System

Section	*ISO 20000 Requirement Keywords*	*CMMI PAs*	*CMMI Practices*	*Confidence*	*Comment*
3	**Requirements for a management system**				
3.1	**Management responsibility**				
	Provide evidence of commitment			0	
3.1 a	Management responsibility: Establish policy			0	
3.1 b	Management responsibility: Communicate importance of objectives			0	
3.1 c	Management responsibility: Ensure customer requirements are determined and met	RD	GP 2.10	30	CMMI is weak in addressing customer satisfaction; special care should be taken to ensure that it is addressed and reviewed.
3.1 d	Management responsibility: Appoint member of management	OPD	GP 2.4	30	
3.1 e	Management responsibility: Allocate resources	IPM	GP 2.3	30	
		OPF	GP 2.3	30	
		PMC	GP 2.3	30	
		PP	GP 2.3	30	
		PPQA	GP 2.3	30	
3.1 f	Management responsibility: Manage risks	RSKM	GP 2.10	60	
3.1 g	Management responsibility: Conduct reviews	IPM	GP 2.10		
		MA	GP 2.10	60	
		PMC	GP 2.10	60	

		PP	GP 2.10	60
		PPQA	GP 2.10	60
		RSKM	GP 2.10	60
3.2	**Documentation requirements**			
	Provide documents and records	OPD	GP 2.9	30
		OPD	SP 1.1	30
3.2 a	Service management documentation: Policies and procedures	CM	GP 2.1	60
		CM	GP 2.2	60
		IPM	GP 2.1	60
		IPM	GP 2.2	60
		PMC	GP 2.1	60
		PMC	GP 2.2	60
		PP	GP 2.1	60
		PP	GP 2.2	60
		RD	GP 2.1	60
		RD	GP 2.2	60
		REQM	GP 2.1	60
		REQM	GP 2.2	60
		RSKM	GP 2.1	60
		RSKM	GP 2.2	60
		SAM	GP 2.1	60
		SAM	GP 2.2	60
		VAL	GP 2.1	60
		VAL	GP 2.2	60

(continued)

Table G.1 ISO 20000:2005 Section 3, Requirements for a Management System (continued)

Section	*ISO 20000 Requirement Keywords*	*CMMI PAs*	*CMMI Practices*	*Confidence*	*Comment*
		VER	GP 2.1	60	
		VER	GP 2.2	60	
3.2 b	Service management documentation: Service level agreements (SLAs)	RD	SP 1.2	60	
		RD	SP 2.1	60	
		REQM	SP 1.2	60	
3.2 c	Service management documentation: Processes and procedures	OPD	SP 1.1	30	OPD addresses development of the OSSP. It is expected that the OSSP will describe the service management processes.
3.2 d	Service management documentation: Maintain records			0	
	Establish procedures and responsibilities for documents	CM	GP 2.2	60	
		OPD	SP 1.1	60	
		PP	SP 2.3	60	
		PPQA	SP 2.2	60	
3.3	**Competence, awareness and training**				
	Define service management roles and responsibilities	CM	GP 2.4		
		IPM	GP 2.4	60	
		OT	SP 1.1	60	
		PMC	GP 2.4	60	

	PP	GP 2.4	60
	RD	GP 2.4	60
	REQM	GP 2.4	60
	RSKM	GP 2.4	60
	SAM	GP 2.4	60
	VAL	GP 2.4	60
	VER	GP 2.4	60
Manage staff competencies and training needs	CM	GP 2.5	60
	IPM	GP 2.5	60
	MA	GP 2.5	60
	OID	GP 2.5	60
	OPD	GP 2.5	60
	OPF	GP 2.5	60
	OT	SP 2.1	60
	PMC	SP 1.1	60
	PP	GP 2.5	60
	PP	SP 2.5	60
	RD	GP 2.5	60
	REQM	GP 2.5	60
	RSKM	GP 2.5	60
	SAM	GP 2.5	60
	VAL	GP 2.5	60
	VER	GP 2.5	60
Ensure awareness of relationship of activities to objectives			0

Table G.2 ISO 20000:2005 Section 4, Planning and Implementing Service Management

Section	*ISO 20000 Requirement Keywords*	*CMMI PAs*	*CMMI Practices*	*Confidence*	*Comment*
4	**Planning and implementing service management**				
4.1	**Plan service management (Plan)**				For most practices, "service management" must be specifically addressed.
	Plan service management	PP	SP 2.7	100	
4.1 a	Service management plan: Scope	PP	SP 1.1	60	
4.1 b	Service management plan: Objectives and requirements	PP	SP 1.1	100	
4.1 c	Service management plan: Processes	CM	GP 3.1	60	
		IPM	GP 3.1	60	
		IPM	SP 1.1	60	
		MA	GP 3.1	60	
		PMC	GP 3.1	60	
		PP	GP 3.1	60	
		PPQA	GP 3.1	60	
		RD	GP 3.1	60	
		REQM	GP 3.1	60	
		RSKM	GP 3.1	60	
		SAM	GP 3.1	60	
		VAL	GP 3.1	60	
		VER	GP 3.1	60	
4.1 d	Service management plan: Management roles and responsibilities	CM	GP 2.4	60	

		IPM	GP 2.4	60
		MA	GP 2.4	60
		PMC	GP 2.4	60
		PP	GP 2.4	60
		PPQA	GP 2.4	60
		RD	GP 2.4	60
		REQM	GP 2.4	60
		RSKM	GP 2.4	60
		SAM	GP 2.4	60
		VAL	GP 2.4	60
		VER	GP 2.4	60
4.1 e	Service management plan: Process interfaces	IPM	SP 1.1	60
		RD	SP 2.3	30
4.1 f	Service management plan: Risk management approach	RSKM	SP 1.1	100
		RSKM	SP 1.2	100
		RSKM	SP 1.3	100
4.1 g	Service management plan: Project interfaces	IPM	SP 3.5	30
4.1 h	Service management plan: Resources, facilities, and budget	PP	SP 2.1	100
		PP	SP 2.4	100
4.1 i	Service management plan: Tools	CM	GP 2.3	100
		IPM	GP 2.3	100
		MA	GP 2.3	100
		PMC	GP 2.3	100

(continued)

Table G.2 ISO 20000:2005 Section 4, Planning and Implementing Service Management (continued)

Section	*ISO 20000 Requirement Keywords*	*CMMI PAs*	*CMMI Practices*	*Confidence*	*Comment*
		PP	GP 2.3	100	
		PPQA	GP 2.3	100	
		RD	GP 2.3	100	
		REQM	GP 2.3	100	
		RSKM	GP 2.3	100	
		SAM	GP 2.3	100	
		VAL	GP 2.3	100	
		VER	GP 2.3	100	
4.1 j	Service management plan: Quality management approach	PPQA	GP 2.2	60	
	Management direction and responsibilities	PMC	GP 2.4	100	
		PP	GP 2.2	100	
		PP	GP 2.4	100	
		PP	GP 2.6	100	
		PP	GP 2.7	100	
		PP	GP 2.9	100	
		PP	SP 3.1	100	
		PP	SP 3.3	100	
	Process plan compatibility	IPM	SP 1.4	60	
		PP	SP 3.1	60	
4.2	**Implement service management and provide the services (Do)**				
4.2 a	Implement service management plan: Funds allocation	CM	GP 2.3	100	

		IPM	SP 1.5	100
		IPM	GP 2.3	60
		MA	GP 2.3	100
		PMC	GP 2.3	100
		PP	GP 2.3	100
		PPQA	GP 2.3	100
		RD	GP 2.3	100
		REQM	GP 2.3	100
		RSKM	GP 2.3	100
		SAM	GP 2.3	100
		VAL	GP 2.3	100
		VER	GP 2.3	100
4.2 b	Implement service management plan: Roles and responsibilities allocation	CM	GP 2.4	100
		IPM	GP 2.4	100
		MA	GP 2.4	100
		PMC	GP 2.4	100
		PP	GP 2.4	100
		PPQA	GP 2.4	100
		RD	GP 2.4	100
		REQM	GP 2.4	100
		RSKM	GP 2.4	100
		SAM	GP 2.4	100
		VAL	GP 2.4	100
		VER	GP 2.4	100

(continued)

Table G.2 ISO 20000:2005 Section 4, Planning and Implementing Service Management (continued)

Section	*ISO 20000 Requirement Keywords*	*CMMI PAs*	*CMMI Practices*	*Confidence*	*Comment*
4.2 c	Implement service management plan: Maintain policies, plans, and procedures	CM	GP 2.1	100	
		CM	GP 2.2	100	
		CM	GP 3.1	100	
		IPM	GP 2.1	100	
		IPM	GP 2.2	100	
		IPM	GP 3.1	100	
		MA	GP 2.1	100	
		MA	GP 2.2	100	
		MA	GP 3.1	100	
		PMC	GP 2.1	100	
		PMC	GP 2.2	100	
		PMC	GP 3.1	100	
		PP	GP 2.1	100	
		PP	GP 2.2	100	
		PP	GP 3.1	100	
		PPQA	GP 2.1	100	
		PPQA	GP 2.2	100	
		PPQA	GP 3.1	100	
		RD	GP 2.1	100	
		RD	GP 2.2	100	
		RD	GP 3.1	100	
		REQM	GP 2.1	100	

		REQM	GP 2.2	100
		REQM	GP 3.1	100
		SAM	GP 2.1	100
		SAM	GP 2.2	100
		SAM	GP 3.1	100
		VAL	GP 2.1	100
		VAL	GP 2.2	100
		VAL	GP 3.1	100
		VER	GP 2.1	100
		VER	GP 2.2	100
		VER	GP 3.1	100
4.2 d	Implement service management plan: Manage risks	PMC	SP 1.3	100
		PP	SP 2.2	100
		RSKM	SP 1.1	100
		RSKM	SP 1.2	100
		RSKM	SP 1.3	100
		RSKM	SP 2.1	100
		RSKM	SP 3.1	100
		RSKM	SP 3.2	100
4.2 e	Implement service management plan: Manage recruiting and skills development	IPM	SP 3.2	60
		IPM	SP 3.4	60
4.2 f	Implement service management plan: Manage facilities and budgets	PMC	GP 2.3	100
		PMC	SP 1.1	100

(continued)

Table G.2 ISO 20000:2005 Section 4, Planning and Implementing Service Management (continued)

Section	*ISO 20000 Requirement Keywords*	*CMMI PAs*	*CMMI Practices*	*Confidence*	*Comment*
4.2 g	Implement service management plan: Manage teams	IPM	SP 1.5	30	There is little in the CMMI regarding managing teams other than in IPPD.
		IPM	SP 3.1	30	
		IPM	SP 3.2	30	
		IPM	SP 3.3	30	
		IPM	SP 3.4	30	
		IPM	SP 3.5	30	
4.2 h	Implement service management plan: Report progress	PMC	GP 2.8	100	
		PMC	SP 1.6	100	
		PMC	SP 1.7	100	
4.2 i	Implement service management plan: Coordinate service management processes	IPM	SP 1.5	60	
4.3	**Monitoring, measuring and reviewing (Check)**				
	Apply suitable methods	CM	GP 2.8	60	
		IPM	GP 2.8	60	
		MA	GP 2.8	60	
		MA	All SPs	60	
		PMC	GP 2.8	60	
		PMC	All SPs	60	
		PP	GP 2.8	60	
		PPQA	GP 2.8	60	
		RD	GP 2.8	60	

	REQM	GP 2.8	60	
	RSKM	GP 2.8	60	
	SAM	GP 2.8	60	
	VAL	GP 2.8	60	
	VER	GP 2.8	60	
Demonstrate process ability	CM	GP 2.8	100	
	IPM	GP 2.8	100	
	MA	GP 2.8	100	
	MA	All SPs	100	
	PMC	GP 2.8	100	
	PMC	All SPs	100	
	PP	GP 2.8	100	
	PPQA	GP 2.8	100	
	RD	GP 2.8	100	
	REQM	GP 2.8	100	
	RSKM	GP 2.8	100	
	SAM	GP 2.8	100	
	VAL	GP 2.8	100	
	VER	GP 2.8	100	
Conduct management reviews: Review conformance with plan and requirements	PMC	SP 1.6	30	CMMI doesn't address requirements of ISO 20000.
	PMC	SP 1.7	30	
	RD	GP 2.8	30	
	RD	GP 2.10	30	
Conduct management reviews: Review effectiveness of implementation	RD	GP 2.10	100	
	REQM	GP 2.10	100	

(continued)

Table G.2 ISO 20000:2005 Section 4, Planning and Implementing Service Management (continued)

Section	*ISO 20000 Requirement Keywords*	*CMMI PAs*	*CMMI Practices*	*Confidence*	*Comment*
	Plan audit program	OPF	GP 2.9	100	
		OPF	SP 1.2	100	
		PPQA	GP 2.2	100	
		PPQA	GP 2.9	100	
	Define audit procedure	OPF	SP 1.2	100	
		PPQA	GP 2.2	100	
		PPQA	GP 3.1	100	
	Maintain auditor objectivity and impartiality	PPQA	GP 2.4	100	
	Avoid auditing own work	PPQA	GP 2.4	100	
	Record review objectives, findings, and remedial actions	PPQA	GP 2.9	60	
		PPQA	SP 2.1	100	
		PPQA	SP 2.2	60	
	Communicate findings to relevant parties	PPQA	SP 2.1	100	
4.4	**Continual improvement (Act)**				
4.4.1	**Policy**				
	Publish service improvement policy	OID	GP 2.1	60	
		OPF	GP 2.1	60	
	Remedy noncompliance	CM	GP 2.9	60	Note that compliance to the ISO standard should be specifically required.
		IPM	GP 2.9	60	
		MA	GP 2.9	60	
		PMC	GP 2.9	60	
		PP	GP 2.9	60	

		PPQA	GP 2.9	60
		PPQA	SP 2.1	60
		RD	GP 2.9	60
		REQM	GP 2.9	60
		RSKM	GP 2.9	60
		SAM	GP 2.9	60
		VAL	GP 2.9	60
		VER	GP 2.9	60
	Define roles and responsibilities	OID	GP 2.4	60
		OPF	GP 2.4	60
4.4.2	**Management of improvements**			
	Prioritize suggested service improvements	OID	SP 1.1	60
		OID	SP 1.2	60
		OID	SP 1.4	60
		OID	SP 2.1	60
		OPF	SP 1.2	60
		OPF	SP 1.3	60
	Control the improvement activity	OID	GP 2.2	100
		OID	SP 2.1	100
		OID	SP 2.2	100
		OID	SP 2.3	100
		OPF	GP 2.2	100
		OPF	SP 2.1	100
		OPF	SP 2.2	100
		OPF	SP 3.1	100

(continued)

Table G.2 ISO 20000:2005 Section 4, Planning and Implementing Service Management (continued)

Section	*ISO 20000 Requirement Keywords*	*CMMI PAs*	*CMMI Practices*	*Confidence*	*Comment*
	Establish process for improving processes continually	OID	GP 2.8	100	
		OID	GP 3.1	100	
		OID	SP 2.1	100	
		OID	SP 2.2	100	
		OID	SP 2.3	100	
		OPD	SP 1.1	100	
		OPF	GP 2.8	100	
		OPF	GP 3.1	100	
		OPF	SP 2.1	100	
		OPF	SP 2.2	100	
		OPF	SP 3.1	100	
4.4.2 a	Improve process continually: Improve individual processes	OPF	SP 3.1	100	
4.4.2 b	Improve process continually: Improve organization processes	OID	GP 2.8	100	
		OID	SP 2.2	100	
		OID	SP 2.3	100	
		OPF	GP 2.8	100	
		OPF	SP 2.1	100	
		OPF	SP 2.2	100	
		OPF	SP 3.1	100	
		OPF	SP 3.4	100	

4.4.3	**Activities**			
4.4.3 a	Improve continuously: Collect and analyze data	MA	SP 2.1	60
		MA	SP 2.2	60
		OPD	GP 2.8	60
		OPF	GP 2.8	60
		OPF	SP 1.2	60
4.4.3 b	Improve continuously: Plan and implement improvements	OID	All SPs	100
		OPF	SP 1.3	100
		OPF	SP 2.1	100
		OPF	SP 2.2	100
		OPF	SP 3.1	100
4.4.3 c	Improve continuously: Consult with parties	OID	SP 2.2	100
		OPF	GP 2.7	100
4.4.3 d	Improve continuously: Set improvement targets	OPF	SP 1.1	100
		OPF	SP 2.1	100
		OPP	SP 1.3	100
4.4.3 e	Improve continuously: Consider all relevant inputs	OID	SP 1.1	60
		OPF	SP 1.3	60
4.4.3 f	Improve continuously: Report on service improvements	OID	SP 2.3	60
		OPF	GP 2.8	60

(continued)

Table G.2 ISO 20000:2005 Section 4, Planning and Implementing Service Management (continued)

Section	*ISO 20000 Requirement Keywords*	*CMMI PAs*	*CMMI Practices*	*Confidence*	*Comment*
4.4.3 g	Improve continuously: Revise service management artifacts when necessary	OID	GP 2.1	60	
		OID	GP 2.2	60	
		OID	SP 2.1	60	
		OID	SP 2.2	60	
		OPD	SP 1.1	60	
		OPF	GP 2.1	60	
		OPF	GP 2.2	60	
		OPF	SP 3.4	60	
4.4.3 h	Improve continuously: Deliver and evaluate approved actions	OID	GP 2.8	100	
		OID	SP 2.2	100	
		OID	SP 2.3	100	
		OPF	GP 2.8	100	
		OPF	SP 3.4	100	

Table G.3 ISO 20000:2005 Section 5, Planning and Implementing New or Changed Services

Section	*ISO 20000 Requirement Keywords*	*CMMI PAs*	*CMMI Practices*	*Confidence*	*Comment*
5	**Planning and implementing new or changed services**				
	Service proposal considerations	CM	SP 2.1	60	
		RD	SP 2.1	60	
		RD	SP 3.3	60	
		RD	SP 3.4	60	
		REQM	SP 1.3	60	
	Use change management process for new or changed services	CM	GP 2.2	60	
		CM	SP 2.1	60	
		CM	SP 2.2	60	
	Funding and resources adequacy	CM	GP 2.3	100	
		IPM	GP 2.3	100	
		MA	GP 2.3	100	
		PMC	GP 2.3	100	
		PP	GP 2.3	100	
		PP	SP 1.4	100	
		PP	SP 2.1	100	
		PP	SP 3.2	100	
		PPQA	GP 2.3	100	
		RD	GP 2.3	100	
		REQM	GP 2.3	100	

(continued)

Table G.3 ISO 20000:2005 Section 5, Planning and Implementing New or Changed Services (continued)

Section	*ISO 20000 Requirement Keywords*	*CMMI PAs*	*CMMI Practices*	*Confidence*	*Comment*
		RSKM	GP 2.3	100	
		SAM	GP 2.3	100	
		VAL	GP 2.3	100	
		VER	GP 2.3	100	
5 a	Plan new or changed services: Roles and responsibilities	CM	GP 2.4	100	
		IPM	SP 3.5	100	
		MA	GP 2.4	100	
		PMC	GP 2.4	100	
		PP	GP 2.4	100	
		PPQA	GP 2.4	100	
		RD	GP 2.4	100	
		REQM	GP 2.4	100	
		RSKM	GP 2.4	100	
		SAM	GP 2.4	100	
		VAL	GP 2.4	100	
		VER	GP 2.4	100	
5 b	Plan new or changed services: Changes to plan	PP	GP 3.1	60	
		PP	SP 1.1	60	
5 c	Plan new or changed services: Communication	PP	SP 2.6	100	
		PP	SP 3.3	100	

5 d	Plan new or changed services: New or changed contracts	SAM	SP 1.3	100
		SAM	SP 2.4	
5 e	Plan new or changed services: Recruitment requirements	PP	SP 2.4	60
5 f	Plan new or changed services: Skills and training requirements	OT	SP 1.1	100
		OT	SP 1.2	100
		OT	SP 1.3	100
		PP	SP 2.5	100
5 g	Plan new or changed services: Processes, measures, methods, and tools	IPM	SP 1.1	60
5 h	Plan new or changed services: Budgets and timescales	PP	SP 1.4	100
		PP	SP 2.1	100
		PP	SP 3.2	100
5 i	Plan new or changed services: Acceptance criteria	VAL	SP 1.3	60
		VER	SP 1.3	60
5 j	Plan new or changed services: Expected outcomes	QPM	SP 1.1	30
		VAL	SP 1.3	100
	Acceptance of new or changed services	VAL	SP 2.1	100
		VAL	SP 2.2	100
		VER	SP 3.1	100
		VER	SP 3.2	100
	Report results of new services			0
	Review results of outcomes			0

Table G.4 ISO 20000:2005 Section 6, Service Delivery Process

Section	*ISO 20000 Requirement Keywords*	*CMMI PAs*	*CMMI Practices*	*Confidence*	*Comment*
6	**Service delivery process**				
6.1	**Service level management**				
	Services agreed to and recorded	RD	GP 2.7	60	
		RD	SP 1.1	60	
		RD	SP 1.2	60	
		REQM	GP 2.7	60	
		REQM	SP 1.1	60	
		REQM	SP 1.2	60	
	Document service in service level agreement(s) (SLAs)	RD	SP 1.1	60	
		RD	SP 2.1	60	
		RD	SP 3.5	60	
	Agree on SLAs and supporting documentation	RD	SP 3.3	60	
		RD	SP 3.4	60	
		RD	SP 3.5	60	
		SAM	GP 2.7	60	
		SAM	SP 1.3	60	
	Exercise change control for SLAs	RD	GP 2.6	60	
		REQM	GP 2.6	60	
		REQM	SP 1.3	60	
	Periodically review SLAs	IPM	SP 2.1	60	
		RD	GP 2.7	60	

		RD	GP 2.8	60
		SAM	SP 2.1	60
	Monitor and report on service levels	MA	SP 2.2	60
		MA	SP 2.4	60
		VER	SP 3.2	60
	Review and report on nonconformance	PMC	SP 2.1	100
		PMC	SP 2.3	100
		VER	SP 3.2	100
	Record actions for improvement	PMC	SP 2.2	100
		VER	GP 3.2	100
		VER	SP 3.2	100
6.2	**Service reporting**			
	Define scope of service report	MA	SP 1.4	60
	Produce quality service reports	MA	GP 2.7	60
		MA	SP 2.4	60
6.2 a	Service reporting: Performance against service level targets	MA	SP 2.2	60
		MA	SP 2.4	60
6.2 b	Service reporting: Noncompliance issues	PMC	SP 2.1	100
		VER	SP 3.2	100
6.2 c	Service reporting: Workload characteristics	MA	SP 2.2	100
		MA	SP 2.4	100
6.2 d	Service reporting: Performance following major events	MA	SP 2.2	100
		MA	SP 2.4	100

(continued)

Table G.4 ISO 20000:2005 Section 6, Service Delivery Process (continued)

Section	*ISO 20000 Requirement Keywords*	*CMMI PAs*	*CMMI Practices*	*Confidence*	*Comment*
6.2 e	Service reporting: Trend information	MA	SP 2.2	100	
		MA	SP 2.4	100	
6.2 f	Service reporting: Satisfaction analysis			0	
	Consider service report for decisions and corrective actions	MA	SP 2.4	100	
6.3	**Service continuity and availability management**				Continuity and availability are not explicitly addressed in CMMI.
	Determine availability and service continuity requirements	RD	SP 1.2	30	
	Include access rights and response times	RD	SP 2.1	30	
		RD	SP 2.2	30	
	Review availability and service continuity plans annually			0	
	Maintain availability and service continuity plans			0	
	Retest availability and service continuity plans			0	
	Assess impact of change	CM	SP 2.1	30	
		CM	SP 2.2	30	
	Measure availability	MA	SP 1.2	60	
		MA	SP 2.1	60	
		MA	SP 2.2	60	
		MA	SP 2.4	60	

	Investigate nonavailability			0
	Keep service continuity plans available			0
	Service continuity plans address return to normal conditions			0
	Test service continuity plan			0
	Record continuity test results			0
6.4	**Budgeting and accounting for IT services**			
6.4 a	Establish policies and processes: Budgeting and accounting	CM	GP 2.1	60
		CM	GP 3.1	60
		IPM	SP 1.1	60
		PMC	GP 2.1	60
		PMC	GP 3.1	60
		PP	GP 2.1	60
		PP	GP 3.1	60
		PP	SP 1.4	60
		PP	SP 2.1	60
		REQM	GP 2.1	60
		REQM	GP 3.1	60
6.4 b	Establish policies and processes: Apportioning indirect and direct costs	CM	GP 2.1	60
		CM	GP 3.1	60
		PP	GP 2.1	60
		PP	GP 3.1	60
		PP	SP 1.1	100
		PP	SP 1.4	100

(continued)

Table G.4 ISO 20000:2005 Section 6, Service Delivery Process (continued)

Section	*ISO 20000 Requirement Keywords*	*CMMI PAs*	*CMMI Practices*	*Confidence*	*Comment*
		PP	SP 2.1	100	
		REQM	GP 2.1	60	
		REQM	GP 3.1	60	
6.4 c	Establish policies and processes: Financial control of processes	CM	GP 2.1	60	
		CM	GP 3.1	60	
		PMC	GP 2.1	60	
		PMC	GP 3.1	60	
		PP	GP 2.1	60	
		PP	GP 3.1	60	
		PP	SP 2.1	100	
		PP	SP 3.2	100	
		PP	SP 3.3	100	
		REQM	GP 2.1	60	
		REQM	GP 3.1	60	
	Budget costs	PP	SP 2.1	100	
		PP	SP 3.2	100	
	Report costs	PMC	SP 1.1	30	
	Apply change management process for changes to services	CM	SP 2.1	100	
		REQM	SP 1.3	100	
6.5	**Capacity management**				"Capacity" is not in CMMI but can be interpreted appropriately in PP, REQM, and MA.

	Develop capacity plan	PP	SP 2.4	30	
6.5 a	Manage capacity: Current and predicted capacity and performance requirements	PP	SP 2.4	30	
6.5 b	Manage capacity: Identify timescales, thresholds, and costs for service upgrades	PP	SP 2.4	30	
6.5 c	Manage capacity: Evaluate effects of anticipated service upgrades	PP	SP 2.4	30	
		REQM	SP 1.1	30	
6.5 d	Manage capacity: Predict impact of external changes	REQM	SP 1.3	30	
6.5 e	Manage capacity: Conduct predictive analysis	MA	SP 2.2	30	
		QPM	SP 2.2	30	
		QPM	SP 2.3	30	
	Identify methods, procedures, and techniques	MA	SP 1.1	30	
		MA	SP 1.2	30	
		MA	SP 1.3	30	
		MA	SP 1.4	30	
6.6	Information security management				Information security is not explicitly addressed in CMMI.
	Approve information security policy			0	
	Communicate information security policy			0	
6.6 a	Establish security controls: Implement policy requirements			0	
6.6 b	Establish security controls: Manage risks	PMC	SP 1.3	30	
		PP	SP 2.2	30	
		RSKM	All SPs	30	

(continued)

Table G.4 ISO 20000:2005 Section 6, Service Delivery Process (continued)

Section	*ISO 20000 Requirement Keywords*	*CMMI PAs*	*CMMI Practices*	*Confidence*	*Comment*
	Document security controls			0	
	Describe risks			0	
	Security controls impact assessment	CM	SP 2.2	30	
		REQM	SP 1.1	30	
		REQM	SP 1.3	30	
	Follow formal agreements with external organizations			0	
	Security incident reporting	PMC	SP 2.1	30	
	Establish procedure for incident investigation			0	
	Establish security incident monitoring mechanisms	MA	SP 1.1	30	
		MA	SP 1.2	30	
		MA	SP 1.4	30	
	Identify improvement actions based on service incidents	MA	GP 3.2	30	

Table G.5 ISO 20000:2005 Section 7, Relationship Processes

Section	*ISO 20000 Requirement Keywords*	*CMMI PAs*	*CMMI Practices*	*Confidence*	*Comment*
7	**Relationship processes**				
7.1	**General**				
7.2	**Business relationship management**				
	Identify service stakeholders/customers	PP	SP 2.6	100	
	Conduct service review at least annually	PMC	SP 1.7	60	
		SAM	SP 1.3	60	
		SAM	SP 2.1	60	
	Conduct interim service review meetings	IPM	SP 2.2	100	
		PMC	SP 1.7	100	
	Document meetings	IPM	SP 2.1	100	
		IPM	SP 2.2	100	
		PMC	SP 1.7	100	
	Revise SLAs	IPM	SP 2.1	100	
		IPM	SP 2.2	100	
		IPM	SP 2.3	100	
		PMC	SP 1.7	100	
		REQM	SP 1.3	100	
		SAM	SP 1.3	100	
		SAM	SP 2.1	100	
	Use change management process for SLA changes	CM	SP 2.1	100	
		CM	SP 2.2	100	
		REQM	SP 1.3	100	

(continued)

Table G.5 ISO 20000:2005 Section 7, Relationship Processes (continued)

Section	*ISO 20000 Requirement Keywords*	*CMMI PAs*	*CMMI Practices*	*Confidence*	*Comment*
	Maintain awareness of business needs	OPP	SP 1.3	30	CMMI emphasis is on quality and process performance (based on business needs).
		QPM	SP 1.1	30	
		RD	SP 1.1	60	
		RD	SP 1.2	60	
		SAM	SP 2.1	60	
	Establish complaints process			0	
	Establish formal service complaint definition	IPM	SP 2.3	30	
	Resolve formal service complaints			0	
	Provide customer complaint escalation process			0	
	Establish customer satisfaction/business relationship responsibility			0	
	Establish process for responding to customer satisfaction measurements			0	
	Identify improvement actions based on service complaints			0	
7.3	**Supplier management**				
	Document supplier management process	OPD	SP 1.1	100	
		SAM	GP 2.2	100	
	Establish supplier contract manager	SAM	GP 2.4	100	
	Document supplier agreement and obtain agreement(s)	SAM	SP 1.3	60	
	Align business and supplier SLAs			0	
	Develop lead/supplier interface agreements	IPM	SP 2.3		

	PI	SP 2.1	100
	PI	SP 2.2	100
	RD	SP 2.3	100
	SAM	SP 1.3	100
	TS	SP 2.3	100
Document lead/supplier roles and responsibilities	SAM	SP 1.3	100
Establish processes for ensuring supplier(s) meet requirements	SAM	SP 2.2	100
Conduct major review of contract/formal agreement at least annually	PMC	SP 1.7	60
	SAM	SP 1.3	60
	SAM	SP 2.1	60
Change contract/SLA(s), as appropriate	REQM	SP 1.3	60
	SAM	SP 1.3	60
	SAM	SP 2.1	60
Use change management process for any changes	CM	SP 2.1	100
	CM	SP 2.2	100
Establish contractual disputes process	IPM	SP 2.3	30
	SAM	SP 1.3	60
	SAM	SP 2.3	30
Establish expected end/early end/transfer process	SAM	SP 1.3	30
Monitor and review performance	PMC	SP 1.7	60
	SAM	SP 2.1	60
Record supplier management improvement actions	PMC	SP 2.2	100
	SAM	SP 2.1	100

Table G.6 ISO 20000:2005 Section 8, Resolution Processes

Section	*ISO 20000 Requirement Keywords*	*CMMI PAs*	*CMMI Practices*	*Confidence*	*Comment*
8	**Resolution processes**				
8.1	**Background**				
8.2	**Incident management**				
	Record incidents			0	
	Establish procedures to manage the impact of incidents			0	
	Establish procedures for resolving incidents			0	
	Inform customer about incident or service request progress			0	
	Provide incident information access			0	
	Apply process for classifying and managing major incidents			0	
8.3	**Problem management**				
	Record problems	CAR	SP 2.3	60	
		MA	SP 2.1	60	

Establish procedures for handling incidents and problems	CAR	SP 1.2	60
Ensure procedures address handling problem from recording through closure	CAR	SP 1.2	60
Take preventive action for potential problems	CAR	SP 2.1	60
Use change management process to handle required changes	CM		
	CM	SP 2.1	60
		SP 2.2	60
Monitor problem resolution for effectiveness	CAR	GP 2.8	60
	CAR	SP 2.2	60
Up-to-date information available to incident management			0
Identify improvement actions based on problem management process	CAR	SP 1.2	100
	CAR	SP 2.3	100
	OID	SP 2.1	100

Table G.7 ISO 20000:2005 Section 9, Control Processes

Section	*ISO 20000 Requirement Keywords*	*CMMI PAs*	*CMMI Practices*	*Confidence*	*Comment*
9	**Control processes**				
9.1	**Configuration management**				
	Establish an integrated approach for change and configuration management	CM	GP 2.2	60	
		CM	SP 1.2	60	
	Establish interface to financial asset accounting processes			0	
	Establish configuration item policy	CM	GP 2.1	100	
		CM	SP 1.1	100	
	Identify configuration item information, relationships, and documentation needed	CM	SP 1.1		
		CM	SP 3.1	60	
	Establish mechanisms for configuration management	CM	SP 1.2	100	
	Ensure sufficient degree of control	CM	GP 2.9	100	
		CM	SP 2.2	100	
	Provide configuration management information to change management process	CM	SP 2.2	100	
	Establish traceability and auditability of changes	CM	SP 3.1	100	
		CM	SP 3.2	100	
	Maintain integrity of components in the configuration control procedures	CM	SP 2.2	100	
		CM	SP 3.2	100	
	Establish baseline of configuration items	CM	SP 1.3	100	
	Control master copies of digital configuration items	CM	SP 1.2	100	

	Uniquely identify and control configuration items	CM	SP 1.1	100
		CM	SP 1.2	100
	Active management and verification of CMDB	CM	SP 1.2	60
	Provide access to configuration information	CM	SP 3.1	100
	Establish configuration audit procedures	CM	SP 3.2	60
9.2	**Change management**			
	Define and document scope of service and infrastructure changes	CM	SP 2.1	60
	Document and classify requests for change			0
	Assess requests for change	CM	SP 2.1	60
	Reverse/remedy unsuccessful change			0
	Approval and check changes	CM	SP 2.1	100
	Implement changes in a controlled manner	CM	SP 2.1	100
		CM	SP 2.2	100
	Conduct review after change implementation	CM	SP 2.2	30
	Establish emergency change policies and procedures			0
	Use scheduled implementation dates of changes as basis for release scheduling	CM	SP 2.2	30
	Maintain change and release schedule and communicate to relevant parties	CM	SP 2.2	30
	Analyze change records			0
	Record results of change record analyses			0
	Identify improvement actions based on change management process	CM	GP 2.8	100
		CM	GP 3.2	100
		OPF	SP 1.3	100
		OPF	SP 2.1	100

Table G.8 ISO 20000:2005 Section 10, Release Process

Section	*ISO 20000 Requirement Keywords*	*CMMI PAs*	*CMMI Practices*	*Confidence*	*Comment*
10	**Release process**				
10.1	**Release management process**				
	Document release policy			0	
	Engage service provider with the business in planning releases	CM	GP 2.2	30	
		CM	SP 2.1	30	
	Obtain agreement and authorization for release plan	CM	GP 2.2	100	
		CM	GP 2.7	100	
	Reverse/remedy unsuccessful release	CM	GP 2.2	30	
		CM	GP 2.7	30	
	Develop the plan for release of services, systems, software, and hardware	CM	GP 2.2	30	
	Release management process	OPD	SP 1.1	30	

Assess impact of change request(s) on release plan	CM	SP 2.1	60
	CM	SP 2.2	60
Include changing of configuration information and change records in release procedures	CM	GP 2.2	100
	CM	SP 2.2	100
Establish process for managing emergency releases			0
Establish controlled acceptance test environment	VAL	SP 1.2	
	VER	SP 1.2	100
Maintain integrity of hardware and software during release and distribution			0
Measure release successes and failures			0
Include postrelease period in measures			0
Analyze release incidents			0
Identify improvement actions based on release management process			0

Index

E

F

G

H

I

L

M

P

Q

S